NATIONAL AERONAUTICS AND SPACE ADMINIST,RATION

NASA

SUPPORT MANUAL

APOLLO
SPACECRAFT FAMILIARIZATION

Contract NAS9-150
Exhibit I; Paragraph 10.2

SID 62-435

1 DECEMBER 1966

MANNED SPACECRAFT CENTER
HOUSTON, TEXAS

Published by Books Express Publishing
Copyright © Books Express, 2012
ISBN 978-1-83931-011-9

Books Express publications are available from all good retail and online booksellers. For publishing proposals and direct ordering please contact us at: info@books-express.com

TABLE OF CONTENTS

Section	Title	Page

LIST OF ILLUSTRATIONS

LIST OF EFFECTIVE PAGES

NOTE: The portion of the text affected by the current changes is indicated by a vertical line in the outer margins of the page.

TOTAL NUMBER OF PAGES IN THIS PUBLICATION IS 190, CONSISTING OF THE FOLLOWING:

Page No.

Title
A
i thru viii
1-1 thru 1-4
2-1 thru 2-14
3-1 thru 3-60
4-1 thru 4-8
5-1 thru 5-8
6-1 thru 6-4
7-1 thru 7-34
8-1 thru 8-28
A-1 thru A-2
B-1 thru B-18

*The asterisk indicates pages changed, added, or deleted by the current change.

Manuals will be distributed as directed by the NASA Apollo Program Office. All requests for manuals should be directed to the NASA Apollo Spacecraft Program Office at Houston, Texas.

NASA

SUPPORT MANUAL

APOLLO
SPACECRAFT FAMILIARIZATION

Contract NAS9-150
Exhibit I; Paragraph 10.2

PREPARED BY NORTH AMERICAN AVIATION, INC.
SPACE AND INFORMATION SYSTEMS DIVISION
TRAINING AND SUPPORT DOCUMENTATION
DEPARTMENT 671

THIS MANUAL REPLACES SM2A-02 DATED 1 DECEMBER 1965

PUBLISHED UNDER AUTHORITY OF THE NATIONAL AERONAUTICS AND SPACE ADMINISTRATION

V-ā

INTRODUCTION

This manual provides general introductory data for personnel associated with the Apollo program. Each command and service module system is discussed in general terms, but with sufficient detail to convey a clear understanding of the systems. In addition, the Apollo earth orbit and lunar landing missions are described, planned, completed, and test programs or missions are identified. Manufacturing, training equipment, ground support equipment, space vehicles, and the lunar module are all covered in gross terms. The source information used in the preparation of this manual was that available as of November 1, 1966.

This manual was prepared for the National Aeronautics and Space Administration by Space and Information Systems Division of North American Aviation, Inc., Downey, California.

A0051

SM-2A-486

PROJECT APOLLO

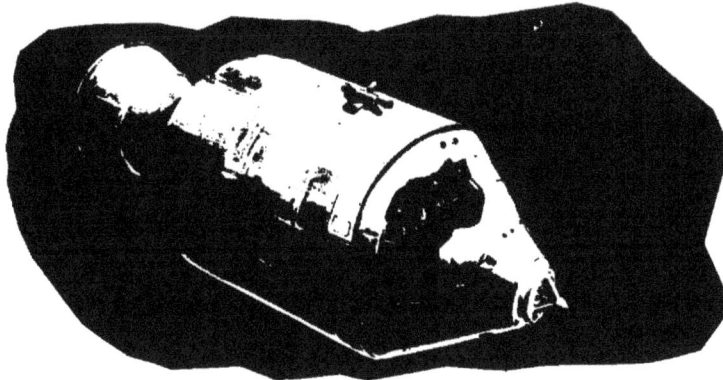

Figure 1-1.

Apollo Spacecraft

SM-2A-1H

1-1. The ultimate objective of Project Apollo is to land men on the moon for limited observation and exploration in the vicinity of the landing area and assure their subsequent safe return to earth. (See figure 1-1.) This objective will climax a series of earth suborbital and orbital missions. Although each of these missions will have specific objectives, they will be flown primarily for state-of-the-art advancement and qualification of systems for the ultimate lunar landing mission.

1-2. The project consists of three phases designed to obtain these ultimate goals:

- The first phase consisted of a number of boilerplate missions for research and developmental purposes. Boilerplates were preproduction spacecraft similar to their production counterparts in shape, size, mass, and center of gravity.

- Phase two is being conducted with limited production spacecraft. These spacecraft are being utilized for systems development and qualification in an earth orbital environment with man in the loop and without the lunar module.

- The third and final phase consists of those missions with the lunar module which will culminate in a manned lunar landing.

1-3. THE APOLLO TEST PROGRAM.

1-4. The Apollo test program is designed to confirm the overall structural integrity, systems performance, compatibility, and intermodular compatibility of the three-man spacecraft. The program follows a path of developmental progress from initial structural integrity confirmation to the complex testing of each module and system for reliability and compatibility. Three basic phases are scheduled for spacecraft testing. The first is research and developmental testing conducted to verify the engineering concepts and basic design employed in the Apollo configuration. The second phase is the qualification testing of the spacecraft hardware and components. The third phase of the test program will verify the production spacecraft systems operation and the man-machine compatibility of the spacecraft. A full test program summary is presented in section VII.

1-5. EARTH SUBORBITAL MISSIONS.

1-6. Two unmanned spacecraft earth suborbital missions have been accomplished to evaluate the command module heat shield performance, and the structural compatibility and integrity of the spacecraft and launch vehicle. These missions also served to qualify and confirm compatibility of the spacecraft-launch vehicle combinations.

1-7. The earth suborbital missions aided in the determination of structural loading, systems performance, and separation characteristics of the launch escape system and boost protective cover from the command module, service module from the adapter, and the command module from the service module. Also, the command module adequacy for manned entry from a low earth orbit was evaluated as well as performance of the service module reaction control system ullage maneuver, service propulsion start, and service propulsion system operation.

1-8. An example of a mission profile for one particular earth suborbital mission is presented in figure 1-2.

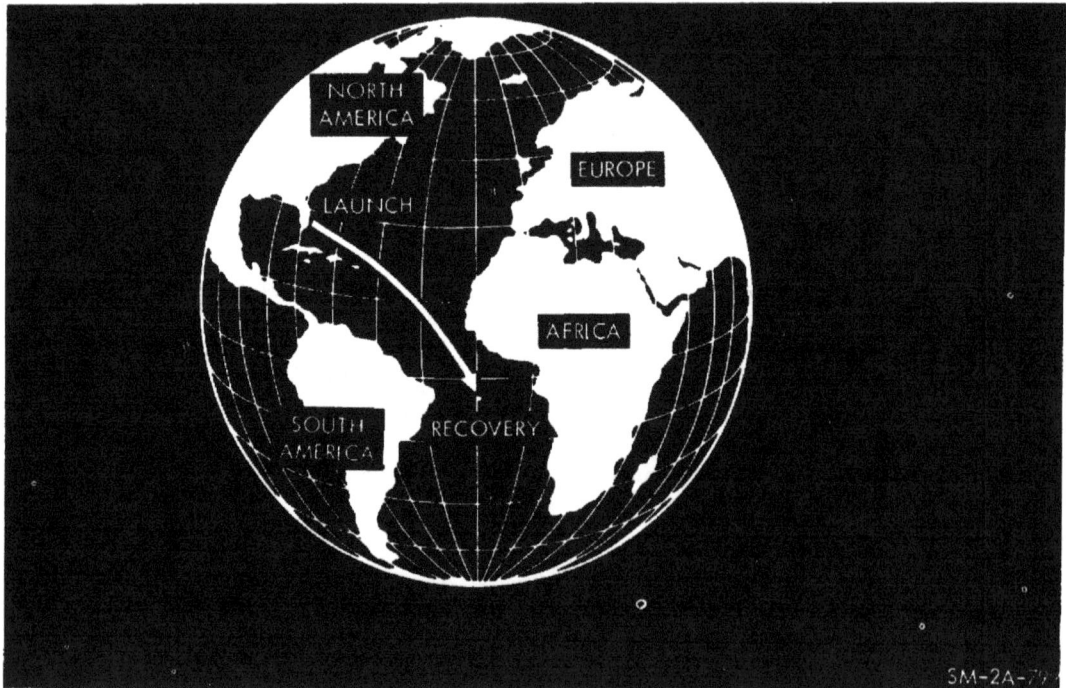

Figure 1-2. Earth Suborbital Mission Profile (Typical)

EARTH ORBITAL MISSION PROFILE

Figure 1-3. Earth Orbital Mission Profile (Typical)

1-9. <u>EARTH ORBITAL MISSIONS.</u>

1-10. Unmanned and manned spacecraft earth orbital missions (figure 1-3) are pro-
grammed to confirm the compatibility of the spacecraft-launch vehicle and spacecraft-
LM combinations to demonstrate spacecraft and lunar module (LM) performance, and
to develop flight crew proficiency.

1-11. The unmanned missions will serve to qualify the launch vehicle and confirm
spacecraft-launch vehicle compatibility. Operating procedures will be analyzed during
these missions to determine the adequacy, feasibility, and overall performance of the
launch vehicle and the unmanned spacecraft.

1-12. Manned missions will be conducted to determine crew and manned space-flight
network (MSFN) proficiency in ascent, earth orbit injection, transposition and docking,
rendezvous and docking, entry, and recovery-phase task requirements. Flight crew
MSFN interface will be constantly conditioned and prepared for deep-space operations.

1-13. The individual mission profile will be determined by the mission objectives for a
given flight. The profiles of earth orbital missions range from circular orbits to
elliptical orbits.

1-14. <u>LUNAR LANDING MISSIONS.</u>

1-15. The lunar landing mission will be accomplished after all other tests and missions
have been satisfactorily completed. The purpose of this mission is to explore the lunar
surface in the vicinity of the LM, and to evaluate the effect of the deep-space environment
upon the flight crew, spacecraft, and the MSFN.

1-16. The lunar landing mission will be of much greater complexity than previous missions. In addition to those tasks required for an earth orbital mission, translunar injection, translunar midcourse corrections, lunar orbit insertion and coast, LM descent, lunar exploration, LM ascent, transearth injection, and transearth midcourse corrections must be accomplished.

1-17. The velocity required for the proper mission profile will be determined by MSC and verified by the Apollo guidance computer of the CSM navigation and control system. After achieving lunar orbit, the flight crew will make observations of a preselected landing site to determine the adequacy of the landing area and/or possible alternate site. Two crew members will then enter the LM through the forward tunnel of the command module, perform a check of the LM systems, and extend the landing gear. At a predetermined point in lunar orbit, the LM will separate from the command and service modules (CSM) and descend to the surface of the moon. The CSM, under control of the remaining crew member, will continue to orbit the moon.

1-18. After landing on the lunar surface, the LM crewmen will alternately egress to the lunar surface and explore the landing site area. During this time, samples of the lunar crust will be taken for subsequent analysis upon return to earth.

1-19. After lunar exploration has been completed, the crew members will re-enter the LM, which will then ascend to rendezvous with the CSM. When docking is completed, the two LM crew members enter the CSM, which is then separated from the LM. Transearth injection (for return to earth), transearth midcourse corrections, entry, and recovery, will then be accomplished.

1-20. The navigation tasks required for the lunar landing mission far exceed those of earth orbital missions. During this mission, the proficiency of flight crew navigation will undergo its severest test of the Apollo program. Figure 1-4 illustrates the typical lunar exploration mission profile with emphasis placed upon the major navigational tasks of the earth-moon relationship. The detailed requirements of the lunar landing (and exploration) mission are described in section VIII.

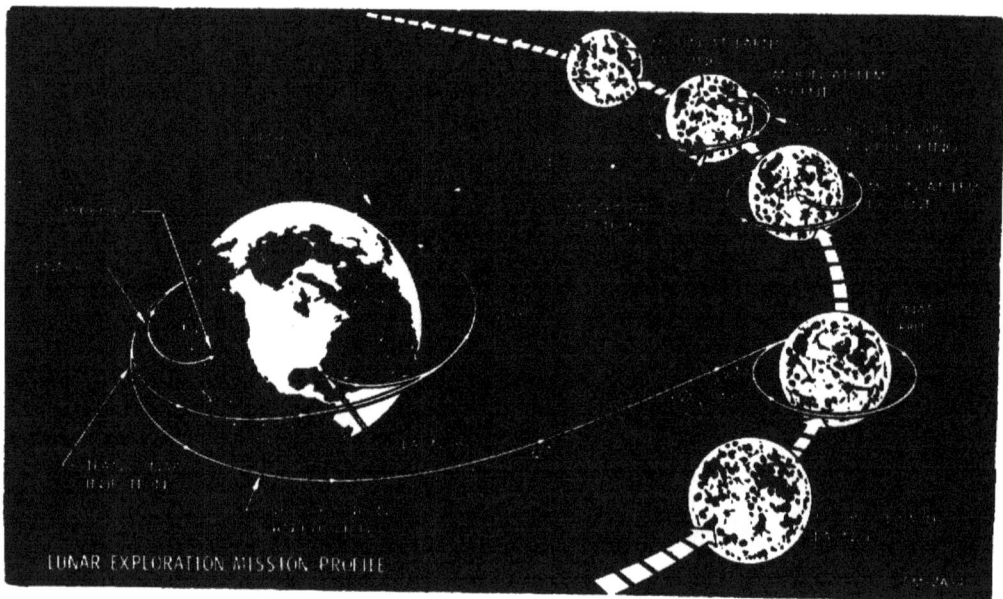

Figure 1-4. Lunar Exploration Mission Profile (Typical)

APOLLO SPACE VEHICLE

2-1. GENERAL.

2-2. The Apollo space vehicles are comprised of various spacecraft and launch vehicle modules. The spacecraft (at launch), based upon mission objectives, may consist of a launch escape system, command module, service module, spacecraft LM adapter (SLA), and lunar module. (See figure 2-1.) The launch vehicle consists of a Saturn booster configuration. The overall height and weight of the space vehicle is directly related to the flight trajectory dictated by mission objectives. Major variances in height and weight are based on the selection of the launch vehicle and configuration of the spacecraft.

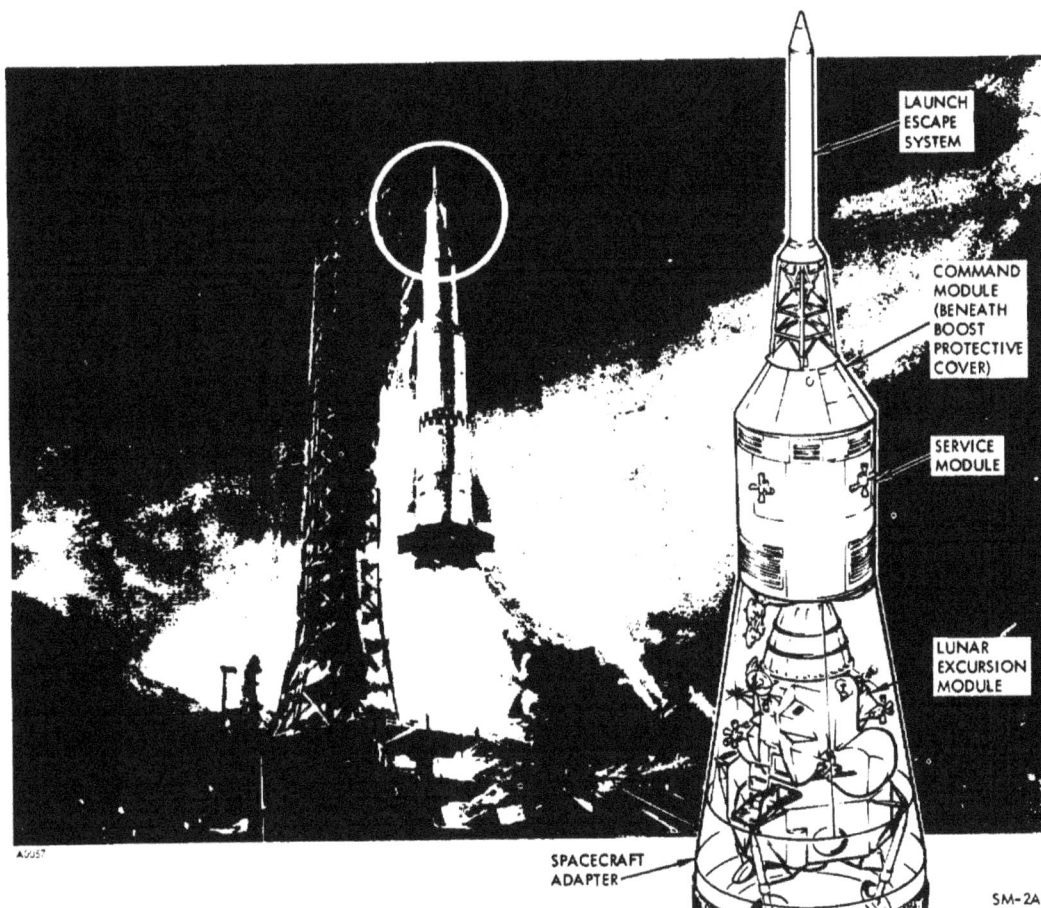

Figure 2-1. Apollo Space Vehicle

2-3.   The external dimensions of the LES, C/M, S/M, and SLA remain constant.   The LM is housed within the SLA and will be installed in the space vehicle for some earth orbital rendezvous and docking missions, and the lunar landing missions.

2-4.   Figure 2-1 depicts the lunar landing space vehicle configuration and geometry of the spacecraft.   The launch vehicle configurations are illustrated later within this section. Refer to the description of launch vehicles for booster configuration variances.

2-5.   APOLLO SPACECRAFT.

2-6.   LAUNCH ESCAPE SYSTEM.

2-7.   The launch escape system provides a means of removing the command module from the space vehicle during a pad abort or atmosphere flight abort.   The launch escape vehicle (figure 2-2) consists of a Q-ball (nose cone), ballast compartment, canard system, three rocket motors, a structural skirt, an open-frame tower, and a boost protective cover.   The structural skirt is secured to the launch escape tower (LET), which transmits stress loads

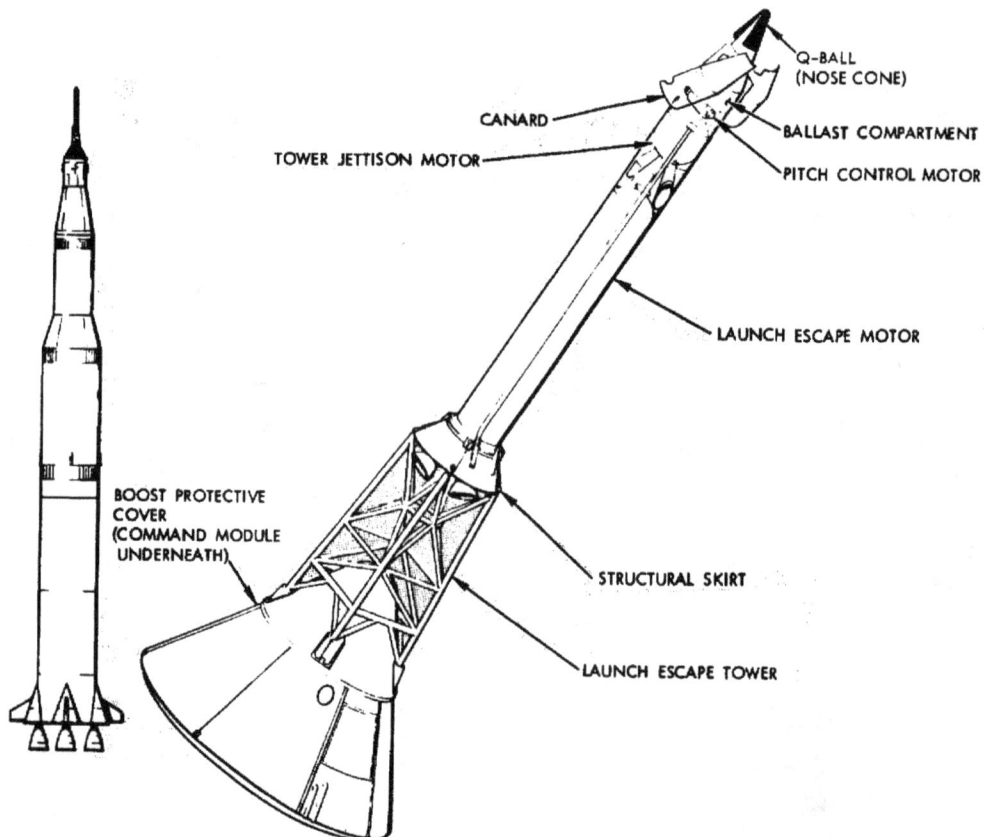

SM-2A-496F

Figure 2-2.   Launch Escape Vehicle

between the launch escape motor and the command module. The boost protective cover (BPC), which protects the C/M exterior during launch and boost, is fastened to the lower end of the tower. Four studs and frangible nuts, one in each tower leg well, secure the tower to the command module structure. After a successful launch or abort mode initiation, explosive detonators fracture the nuts and free the tower. The rocket motors, canards, and explosive squibs are activated by electronic sequencing devices within the CSM. Refer to section III for system operational data.

2-8. COMMAND MODULE.

2-9. The command module (figure 2-3) is the recoverable portion of the spacecraft and, houses the flight crew, the equipment necessary to control and monitor the spacecraft systems, and equipment required for the comfort and safety of the crew. The primary structure of the command module is encompassed by three heat shields, forming a conical-shape exterior. The forward, crew, and aft heat shield structures are coated with ablative material and joined to the primary structure. An insulation material is installed between the primary structure and the heat shields. The C/M consists of three compartments: forward, crew, and aft.

2-10. FORWARD COMPARTMENT. The forward compartment (figure 2-4) is a section between the forward heat shield and the forward side of the forward pressure bulkhead. The center portion is occupied by a tunnel which permits crew members to transfer to the LM and return to the crew compartment during the performance of lunar mission tasks. The interior of the forward compartment is divided into four 90-degree segments which

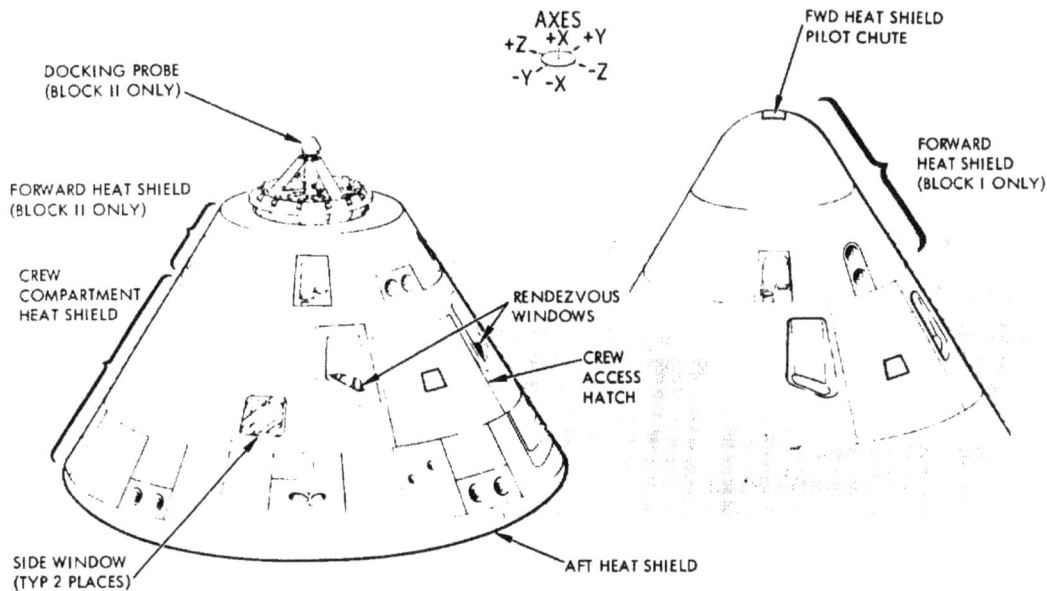

Figure 2-3. Command Module

SM-2A-795A

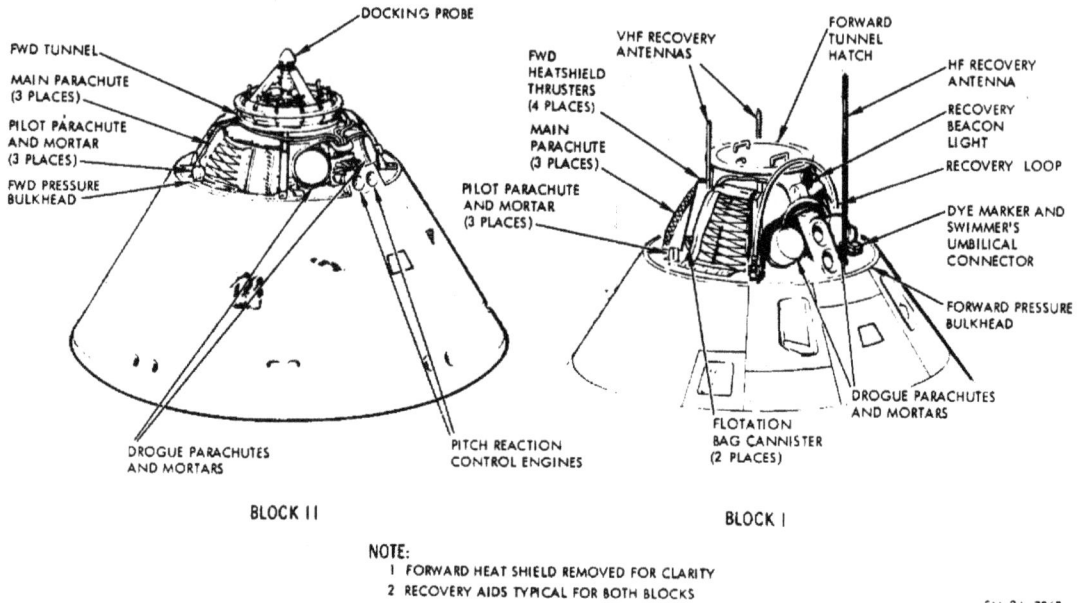

Figure 2-4. Command Module Forward Compartment

contain earth landing system (ELS) components, recovery equipment, two reaction control motors, and the forward heat shield jettisoning mechanism. The major portion of this section contains the active components of the ELS consisting of three main parachutes, three pilot parachutes, two drogue parachutes, as well as drogue and pilot parachute mortars, risers, and the necessary hardware. The recovery equipment, installed in the forward compartment, consists of three flotation bags of the uprighting system, a beacon light, a sea dye marker, a swimmer umbilical, three recovery antennas and a recovery pickup loop. Four thruster-ejectors are installed in the forward compartment to eject the forward heat shield during landing operations. The thrusters operate to produce a rapid, positive release of the heat shield, preventing parachute fabric damage.

2-11. CREW COMPARTMENT. The crew compartment (figure 2-5) is a sealed, three-man cabin with pressurization maintained by the environmental control system. The crew compartment contains the necessary systems, spacecraft controls and displays, observation windows, access hatches, food, water, sanitation, and survival equipment. The compartment incorporates windows and equipment bays as a part of the structure. The following listing contains specific items contained in the crew compartment and their locations. Items marked (Block I) or (Block II), indicate the specific block in which they are included; all others are included in both blocks.

Aft Equipment Storage Bay (See figure 2-5.)

Space suits (two)	TV zoom lens
Space suit spare parts	Portable life support system (Block II)
Umbilicals	CO_2-odor absorber filters
Rest station restraint	Fecal canister
Drogue and probe stowage (Block II)	Helmet storage (two)
Life vests (three)	Communication helmet storage (Block I)

Lower Equipment Bay (See figure 2-5.)

Power servo assembly (PSA)	Display electronics (Block II)
Computer control panel	Reaction jet driver (Block II)
Signal conditioning equipment	Apollo guidance computer (AGC) (Block I)
Rate gyro assembly (RGA) (Block II)	CMC Command Module Computer (Block II)
Electronic control assemblies (five) (Block I)	Medical supplies
Gas chromatograph (Block I)	Medical refrigerator (Block II)
Workshelf (Block I)	Data distribution panel (Block I)
Pulse-code modulation (PCM) units (two)	Attitude gyro-accelerometer assembly (AGAA)
S-band power amplifier	Crew flight data file (CFDF)
Unified S-band equipment and spares	VHF multiplexer (Block I)
Junction box	C-band transponder (Block I)
Motor switches (three)	Audio center equipment
Battery charger	Central timing equipment (CTE)
Sextant and scanning telescope	Entry batteries (three)
Rendezvous radar control (Block II)	Circuit breaker panel
TVC servo amplifier (Block II)	Guidance and navigation (G&N) control panel
Control electronics (Block II)	Coupling display unit (CDU) (Block II)
Gyro display (Block II)	Data storage equipment

Food storage

Scientific equipment

Flight qualification recorder (Block I)

Up-data link

Premodulation processor

VHF/AM transceiver and VHF recovery
beacon

Triplexer (Block II)

VHF/FM transmitter and HF transceiver

R-F switch

Inverters (three)

A-C power box

Pyrotechnic batteries (two)

Lighting control (Block II)

Clock and event timer (Block II)

In-flight tool set

Left-Hand Forward Equipment Bay (See figure 2-5.)

Cabin air fan

Optical storage (Block I)

Translation controller connector panel

Loose parts storage (Block I)

Pressure suit connectors (three)

Water delivery assembly

Food reconstitution device

Clothing (storage)

Cabin heat exchanger

Clock and event-timer panel (Block I)

Thermal radiation coverall (storage)
(Block I)

Radiation survey meter (Block II)

Left-Hand Equipment Bay (See figure 2-5.)

Cabin pressure relief valve

Fixed shock attenuation panel

Environmental control system (ECS)
water and oxygen control panels

Surge tank

Environmental control unit

Pressure hatch stowage (Block II)

CO_2 sensor

Removable shock attenuation panels
(two)

Right-Hand Forward Equipment Bay (See figure 2-5.)

Individual survival kits (three on Block I
S/C) (two on Block II S/C)

System test equipment (Block I)

Waste storage inlet

TV camera and mount food (Block I)

Optical storage (Block II)

Medical supplies

LM docking target storage (Block II)

Bio-instrument accessories (Block II)

Sanitary supplies

Tools and belt (Block II)

FORWARD
COMPARTMENT

CREW COMPARTMENT

CREW COUCH
(TYPICAL)

LEFT-HAND FORWARD
EQUIPMENT BAY

RIGHT-HAND
FORWARD
EQUIPMENT BAY

LOWER
EQUIPMENT BAY

AXES
+X
+Z   +Y
-Y   -Z
-X

FORWARD ACCESS HATCH

FORWARD
COMPARTMENT

CREW
COMPARTMENT

RIGHT-HAND
EQUIPMENT BAY

AFT EQUIPMENT STORAGE BAY

LEFT-HAND EQUIPMENT BAY

AFT COMPARTMENT

AFT COMPARTMENT

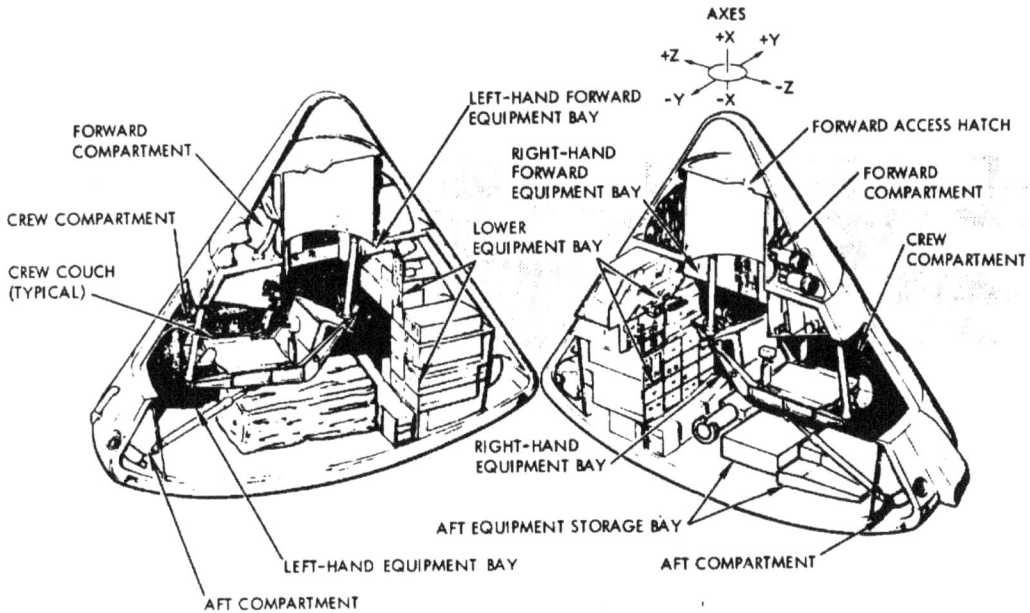

NOTE:
CENTER COUCH REMOVED FOR CLARITY

BLOCK I

AXES
-Y   +X
+Z
-Z
-X   +Y

FORWARD
COMPARTMENT

CREW
COMPARTMENT

LEFT-HAND FORWARD
EQUIPMENT BAY

RIGHT-HAND FORWARD
EQUIPMENT BAY

LOWER
EQUIPMENT
BAY

FORWARD
COMPARTMENT

RIGHT-HAND EQUIPMENT BAY

LEFT-HAND EQUIPMENT BAY

AFT EQUIPMENT STORAGE BAY

CREW COMPARTMENT

AFT COMPARTMENT

AFT COMPARTMENT

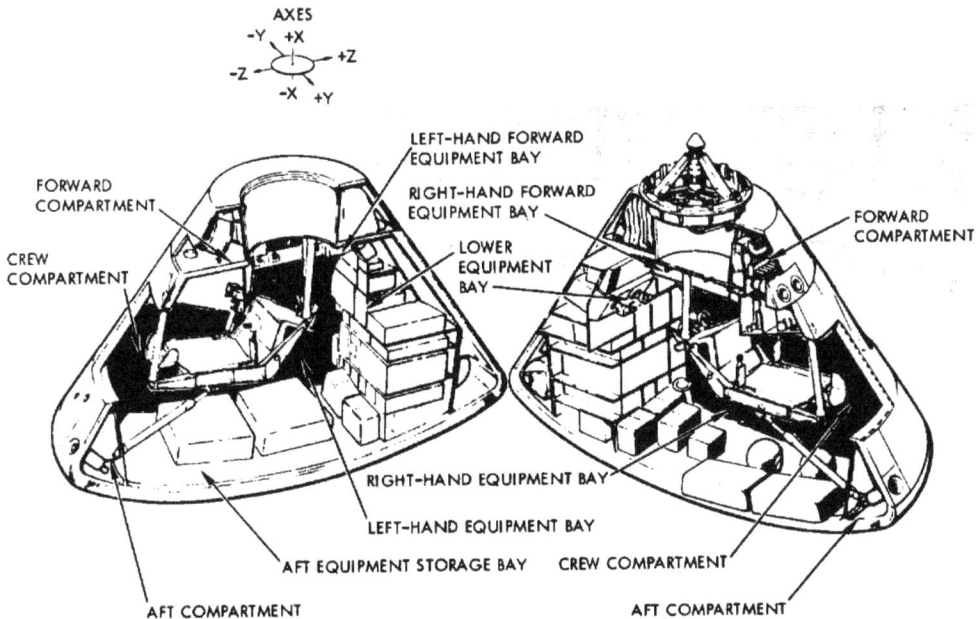

NOTE:
CENTER COUCH OMITTED FOR CLARITY

BLOCK II

SM-2A-498G

Figure 2-5.   Command Module Compartments and Equipment Bays (Typical)

Right-Hand Equipment Bay (See figure 2-5.)

Vacuum cleaner	Fuse box
Electrical power equipment	Food (Block II)
Master Event Sequencers	Waste storage compartment
Power distribution box	ELS sequencers
Circuit utilization box	Master Event Sequencers
Phase correction capacitor box	Signal conditioners
Waste management system controls	Thermal hatch stowage (Block II)

2-12. AFT COMPARTMENT. The aft compartment (figure 2-5) is an area encompassed by the aft portion of the crew compartment heat shield, aft heat shield, and aft sidewall of the primary structure. This compartment contains 10 reaction control motors, impact attenuation structure, instrumentation, and storage tanks for water, fuel, oxidizer, and gaseous helium.

2-13. SERVICE MODULE.

2-14. The service module structure is a cylinder formed by panels of one-inch aluminum honeycomb. (See figure 2-6.) Its interior is unsymmetrically divided into six sectors by radial beams, or webs made from milled aluminum alloy plate. The interior consists of diametrically opposed sectors around a circular section 44 inches in diameter. Sectors 1 and 4 are 50-degree segments and sectors 2 and 5 are 70-degree segments. The remaining sectors, 3 and 6, are 60-degree segments. The equipment contained within the S/M is accessible through maintenance doors located strategically around the exterior surface of the module. The specific items contained in the S/M compartments, and their location, are listed in paragraph 2-16.

2-15. An area between the service and command modules provides space for radial beam trusses connecting these two modules. Beams one, three, and five have compression pads for support of the C/M; beams two, four, and six have compression pads, shear pads, and tension ties. A flat center section in each tension tie, incorporates redundant explosive charges for service module-command module separation. The entire separation system is enclosed within a fairing 26 inches high and 12'-10" in diameter.

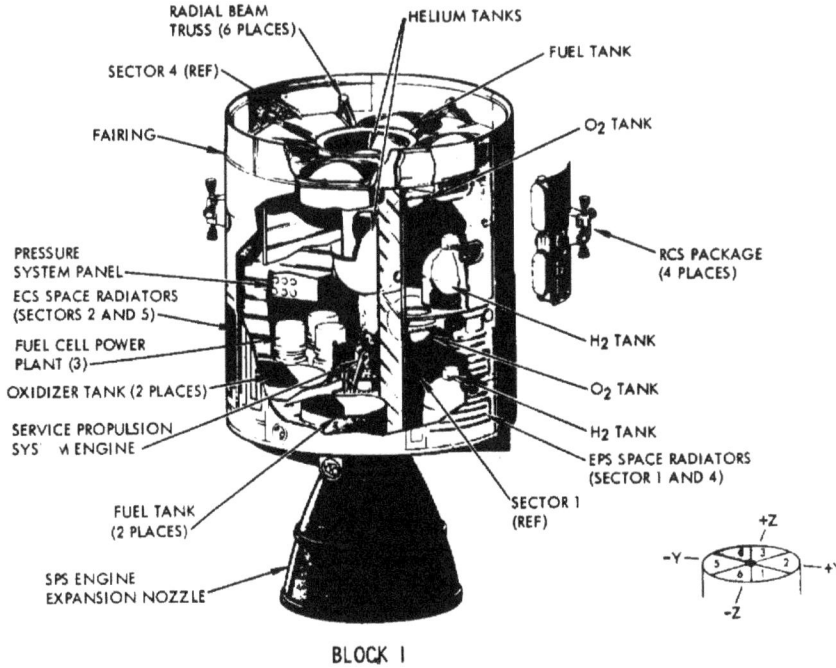

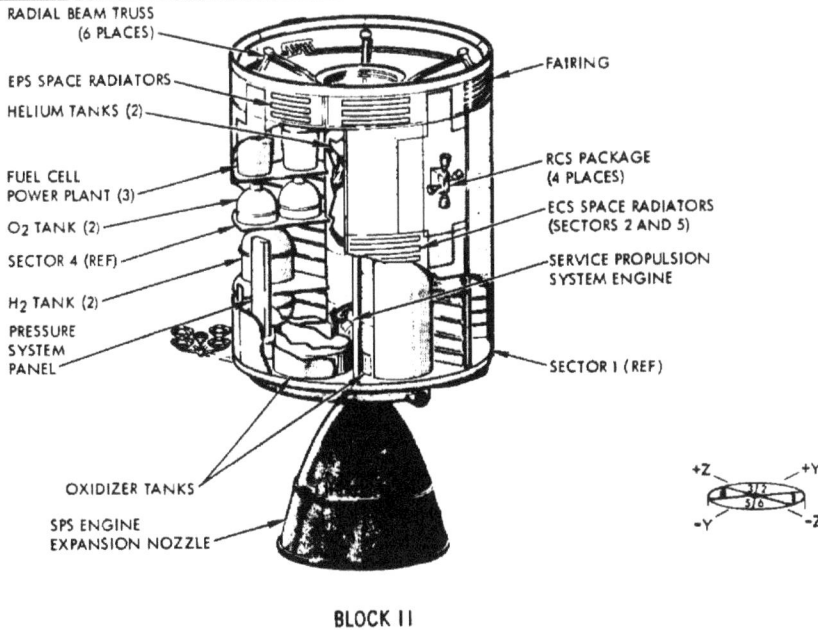

RADIAL BEAM
TRUSS (6 PLACES)

HELIUM TANKS

SECTOR 4 (REF)

FUEL TANK

FAIRING

O₂ TANK

PRESSURE
SYSTEM PANEL

RCS PACKAGE
(4 PLACES)

ECS SPACE RADIATORS
(SECTORS 2 AND 5)

FUEL CELL POWER
PLANT (3)

H₂ TANK

OXIDIZER TANK (2 PLACES)

O₂ TANK

SERVICE PROPULSION
SYSTEM ENGINE

H₂ TANK

EPS SPACE RADIATORS
(SECTOR 1 AND 4)

FUEL TANK
(2 PLACES)

SECTOR 1
(REF)

SPS ENGINE
EXPANSION NOZZLE

+Z

-Y

+Y

-Z

BLOCK I

RADIAL BEAM TRUSS
(6 PLACES)

EPS SPACE RADIATORS

FAIRING

HELIUM TANKS (2)

FUEL CELL
POWER PLANT (3)

RCS PACKAGE
(4 PLACES)

O₂ TANK (2)

ECS SPACE RADIATORS
(SECTORS 2 AND 5)

SECTOR 4 (REF)

SERVICE PROPULSION
SYSTEM ENGINE

H₂ TANK (2)

PRESSURE
SYSTEM
PANEL

SECTOR 1 (REF)

OXIDIZER TANKS

+Z

+Y

SPS ENGINE
EXPANSION NOZZLE

-Y

-Z

BLOCK II

SM-2A-499H

Figure 2-6. Service Module

2-16. Items and their locations contained in Block I and Block II service modules (manned S/C only) are listed as follows.

Location	Contents	
	Block I	Block II
Sector 1	Electrical power system space radiators Cryogenic oxygen tank (two) Cryogenic hydrogen tank (two)	
Sector 2	Environmental control system space radiator Service propulsion system oxidizer tank Reaction control system engine cluster (+Y-axis) Reaction control system helium tank Reaction control system fuel tank Reaction control system oxidizer tank Space radiator isolation valve (two)	Environmental control system space radiator Service propulsion system oxidizer tank Reaction control system engine cluster (+Y-axis) Reaction control system helium tank Reaction control system fuel tank Reaction control system oxidizer tank Space radiator isolation valve (two)
Sector 3	Service propulsion system fuel tank Reaction control system engine (cluster (+Z-axis) Reaction control system helium tank Reaction control system fuel tanks Reaction control system oxidizer tanks Rendezvous radar transponder	Service propulsion system oxidizer tank Reaction control system engine cluster (+Z-axis) Reaction control system helium tank Reaction control system fuel tanks Reaction control system oxidizer tanks Rendezvous radar transponder
Sector 4	Electrical power system space radiator Fuel cell power plant (three) Helium distribution system Reaction control system control unit Electrical power system power control relay box Service module jettison controller (SMJC) sequencer (two)	Fuel cell power plant (three) Cryogenic oxygen tank (two) Cryogenic hydrogen tank (two) Reaction control system control unit Electrical power system power control relay box High-gain antenna (stowed under) Service module jettison controller (SMJC) sequencer (two)
Sector 5	Environmental control system space radiator Service propulsion system oxidizer tank	Environmental control system space radiator Service propulsion system fuel tank

	Contents	
Location	Block I	Block II
Sector 5 (Cont)	Reaction control system engine cluster (-Y-axis) Reaction control system helium tank Reaction control system fuel tank Reaction control system oxidizer tank	Reaction control system engine cluster (-Y-axis) Reaction control system helium tank Reaction control system fuel tanks Reaction control system oxidizer tanks Helium distribution system
Sector 6	Space radiator selection valve (two) Glycol shutoff valve (two) Reaction control system engine cluster (-Z-axis) Reaction control system helium tank Reaction control system fuel tank Reaction control system oxidizer tank Service propulsion system fuel tank	Space radiator selection valve (two) Glycol shutoff valve (two) Reaction control system engine cluster (-Z-axis) Reaction control system helium tank Reaction control system fuel tanks Reaction control system oxidizer tanks Service propulsion system fuel tank
Center section	Service propulsion system helium tank (two) Service propulsion system engine	Service propulsion system helium tank (two) Service propulsion system engine

2-17. LUNAR MODULE.

2-18. The LM, manufactured by Grumman Aircraft Engineering Corp., is a space vehicle which provides a means of transportation for two crewmembers of the Apollo spacecraft to leave the command module, land on the lunar surface, and return to the command module. The LM is then jettisoned from the C/M and left as a lunar satellite. A description of the LM is presented in section IV.

2-19. SPACECRAFT LM ADAPTER.

2-20. The spacecraft LM adapter (figure 2-7) is the structural interstage between the launch vehicle and the spacecraft. The spacecraft LM adapter (SLA) is required on Apollo spacecraft using an uprated Saturn I or Saturn V launch vehicle. (See figure 2-8.) The SLA will house the service propulsion engine expansion nozzle, high-gain antenna, and LM. An umbilical cable is incorporated in the adapter to connect circuits between the launch vehicle and the spacecraft.

2-21. The SLA (figure 2-7) is a tapered cylinder comprised of eight panels, four of which have linear explosive charges installed at panel junctions. During CSM/SLA separation, the charges are fired to open the four panels, free the spacecraft from the launch vehicle, and expose the LM.

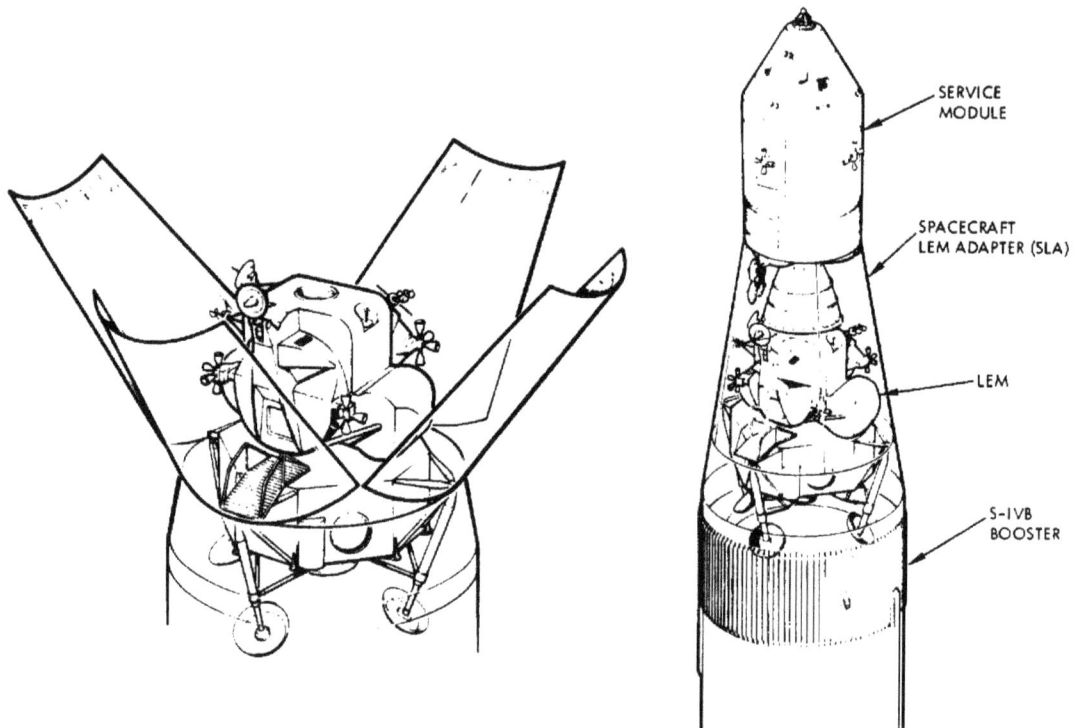

SM-2A-497E

Figure 2-7. Spacecraft LM Adapter

2-22. <u>LAUNCH VEHICLES.</u>

2-23. Launch vehicles used in the Apollo program are illustrated in figure 2-8. The earlier test evaluation and qualification flight vehicles were powered by the launch escape vehicle Little Joe II, Saturn I, and Saturn uprated I launch vehicles. As the Apollo program progresses, the extended lunar mission performance and greater payload necessitates the use of the Saturn V. The general configurations of the launch vehicle boosters are summarized in paragraphs 2-24 through 2-32.

2-24. LAUNCH ESCAPE VEHICLE.

2-25. The launch vehicle used for pad-abort tests (figure 2-2) consisted of a C/M, boost protective cover, launch escape tower, launch escape system motor, a pitch control motor, and a tower jettison motor. Each of the motors used a solid propellant. The LES motor, manufactured by Lockheed Propulsion Corporation, provided up to 155,000 pounds thrust. The pitch control motor, also manufactured by Lockheed Propulsion Corporation, provided up to 3000 pounds thrust. The tower jettison motor, manufactured by Thiokol Chemical Corporation, provided up to 33,000 pounds thrust.

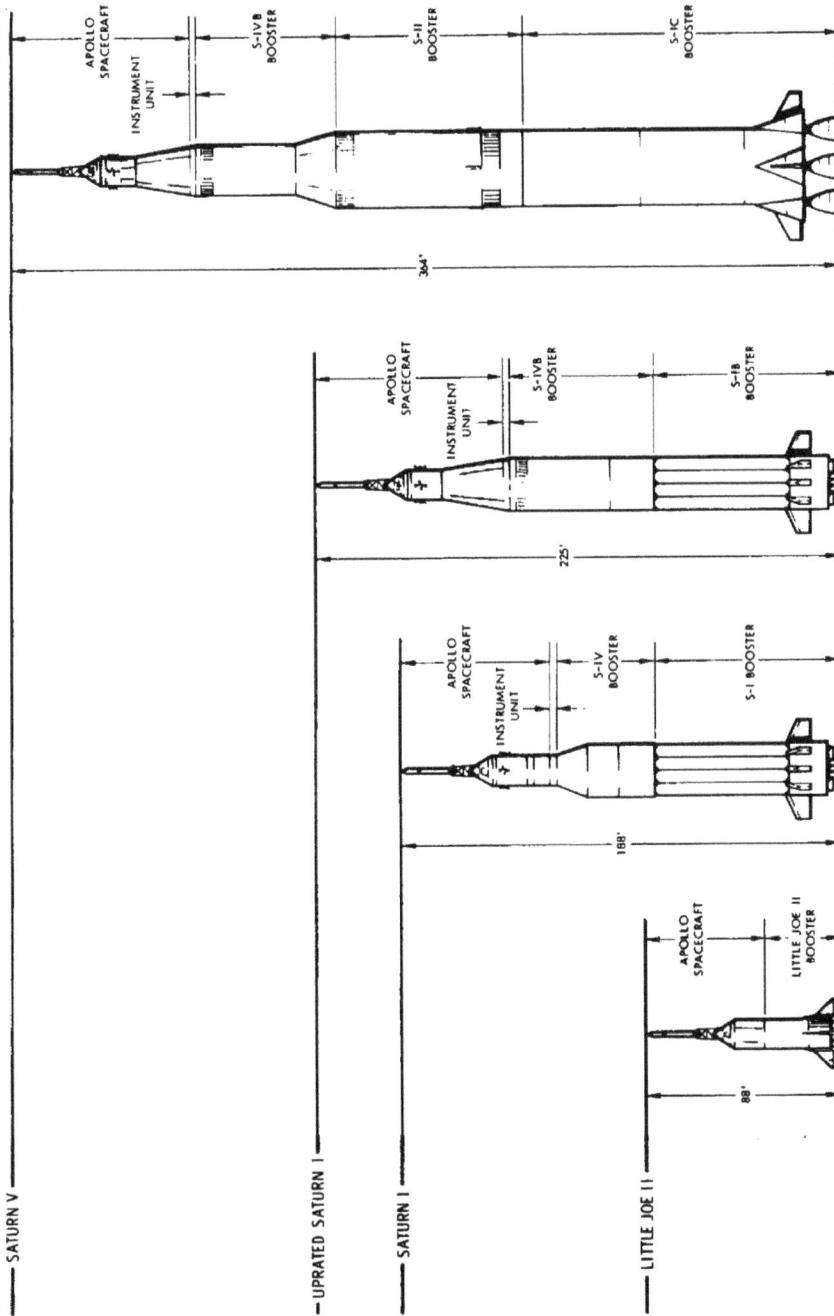

Figure 2-8. Launch Vehicle Configurations

2-26. LITTLE JOE II.

2-27. The Apollo transonic and high-altitude abort tests utilized the Little Joe II launch vehicle. (See figure 2-8.) The launch vehicle was approximately 13 feet in diameter and 29 feet in length. Little Joe II, manufactured by General Dynamics, Convair, used a combination of Algol and Recruit solid-propellant motors. One Algol and six Recruit motors provided an initial boost of 310,000 pounds thrust.

2-28. SATURN I.

2-29. Saturn I consisted of an S-I first-stage booster and an S-IV second stage. (See figure 2-8.) The S-I, manufactured by Chrysler Corporation, was 257 inches in diameter and approximately 82 feet in length. Eight Rocketdyne H-I engines were used, each burning RP-1 and liquid oxygen, and each producing 188,000 pounds thrust. Total boost for the S-I was 1,500,000 pounds thrust. The S-IV, manufactured by Douglas Aircraft Company, was 220 inches in diameter and 40 feet in length. Six Pratt & Whitney RL-10 engines were used, each burning liquid hydrogen and liquid oxygen, and each producing 15,000 pounds thrust. The total boost for the S-IV was 90,000 pounds. An instrument unit, located between the S-IV and the boilerplate adapter, controlled each of the two stages during flight.

2-30. UPRATED SATURN I.

2-31. The uprated Saturn I is a more powerful version of Saturn I, consisting of an S-IB first-stage booster and an S-IVB second stage. (See figure 2-8.) The S-IB, manufactured by Chrysler Corporation, is a lightweight version of the S-I booster, but approximately the same size. A weight reduction of 15,000 pounds per engine, or 120,000 pounds total, is realized while maintaining 1,600,000 pounds thrust. The S-IVB, manufactured by Douglas Aircraft Company, is 260 inches in diameter, 58 feet in length, and of an entirely different configuration than the S-IV. S-IVB employs a single Rocketdyne J-2 engine, burning liquid hydrogen and liquid oxygen, to produce approximately 200,000 pounds thrust. An instrument unit, located between the S-IVB and the SLA, controls each of the two stages during flight.

2-32. SATURN V.

2-33. Saturn V is a three-stage launch vehicle consisting of an S-IC first-stage booster, S-II second stage, and an S-IVB third stage. (See figure 2-8.) The S-IC, manufactured by the Boeing Company, is 33 feet in diameter and 138.5 feet in length; it uses five Rocketdyne F-1 engines. Each F-1 engine, burning RP-1 and liquid oxygen, produces 1,500,000 pounds thrust for an overall boost of 7,500,000 pounds. The S-II, manufactured by the Space and Information Systems Division of North American Aviation, Inc., is 33 feet in diameter and approximately 82 feet in length and employs five Rocketdyne J-2 engines. Each J-2 engine, burning liquid hydrogen and liquid oxygen, produces 200,000 pounds thrust for an overall boost of 1,000,000 pounds. The S-IVB is similar to the second stage of uprated Saturn I, producing 200,000 pounds thrust. An instrument unit, located between the S-IVB and the SLA, controls each of the three stages during flight.

SPACECRAFT SYSTEMS

3-1.   GENERAL.

3-2.   This section contains data relative to the basic nature of the operational command and service module spacecraft systems. The purpose of each system, its functional description, and interface information are presented with a minimum of detailed text consistent with understanding. Concepts are supported by illustrations, listings, and diagrams. Also illustrated are the various panel arrangements within the command module that contain the controls and displays. Data presented covers each complete system as it will be installed in a manned spacecraft for an earth orbital or a lunar landing mission. Systems and components will be tested and qualified prior to manned missions. Other missions, using boilerplates or unmanned S/C, are not covered in this section as these vehicles may contain incomplete or modified systems. Physical differences between Block I and Block II S/C are illustrated in this section. For an explanation of Block I and Block II, refer to section VII.

3-3.   Redundancy is necessary for mission critical items throughout the spacecraft systems in order to maintain the high reliability rates prescribed for the Apollo program. Included are redundant components, power sources, paths for fluids and electrical signals, and redundant operational procedures.

3-4.   LAUNCH ESCAPE SYSTEM.

3-5.   The launch escape system (LES) provides immediate abort capabilities from the launch pad, or away from the path of the launch vehicle in the event of an abort shortly after launch. (See figure 3-1.) Upon abort initiation, the command module will be propelled to a sufficient altitude and lateral distance away from the danger area for the effective operation of the earth landing system. Upon completion of an abort, or a successful launch, the launch escape assembly is jettisoned from the C/M.

3-6. The LES consists of two major structures, plus electrical control equipment located in the C/M. The forward structure is cylindrical, housing three rocket motors (pitch control, tower jettison, launch escape) and a ballast compartment topped by a Q-ball (nose cone). Two canard surfaces are installed below the nose cone. The rocket motors are loaded with solid propellants of various grain patterns, depending upon motor performance requirements. The aft structure is a four-leg, welded, tubular titanium tower. It serves as an intermediate structure, transmitting loads between the C/M and the launch escape assembly, and positioning the C/M a suitable distance from the launch escape motor exhaust. (See figure 3-1.) At its forward end, the tower is attached to a structural skirt that covers the exhaust nozzles of the launch escape motor, and at the aft end, attached to the C/M by four frangible nuts and studs. A boost protective cover is also attached to the aft end of the tower to protect the C/M from the launch escape motor exhaust and boost heating. The LES sequence controllers are located in the C/M and control the system by transmitting signals that ignite the rocket motors, deploy the canard surfaces, and detonate the separation devices. The Q-ball has four static ports for measuring ΔP which is a function of angle of attack. An angle-of-attack indicator is located on the main display console and displays the combined pitch and yaw vectors in terms of percentages.

3-7. LES OPERATION.

3-8. The LES is initiated automatically by the emergency detection system (EDS) of the launch vehicle up to 90 seconds from lift-off, or manually by the astronauts at any time from pad to launch escape tower jettison as shown in figure 3-3. Upon receipt of an abort signal, regardless of its source, the booster is cut off (after the first 40 seconds of flight), and the launch escape systems is activated. Three basic modes for sequencing of the launch escape system are shown in figure 3-3. They are: pad to 30,000 feet, 30,000 feet

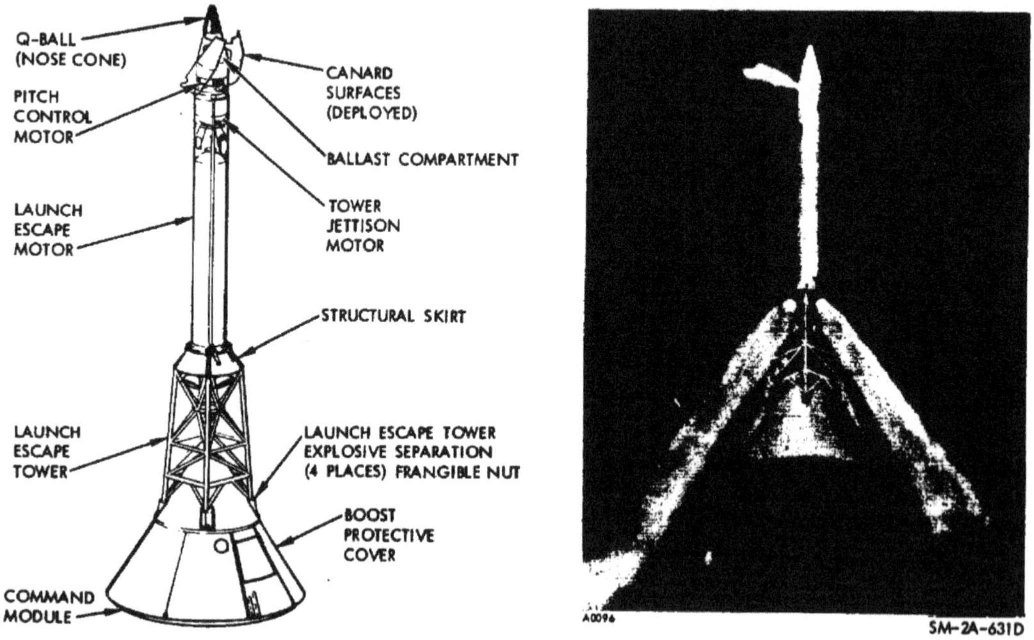

Figure 3-1. Launch Escape Vehicle

to 100,000 feet, and 100,000 feet to tower jettison. Activation of the earth landing system is automatically initiated by the sequencing units of the launch escape system.

3-9. During a successful launch, the launch escape assembly will be jettisoned after reaching a prescribed altitude. The tower explosive devices will be detonated, the jettison motor ignited, and the launch escape assembly (including the boost protective cover) will be propelled away from the path of the spacecraft and booster.

3-10. CANARD OPERATION. The canard (figure 3-2) consists of two deployable surfaces and an operating mechanism. The surfaces are faired into the outer skin of the launch escape assembly below the nose cone and the operating mechanism is inside the cylindrical-shaped assembly. Each surface is mounted on two hinges and is opened by a pyro cylinder that operates the opening mechanism. The pyro cylinder piston is normally in the extended position with the canard surfaces closed. Eleven seconds after an abort signal is received by the launch escape system, an electrical current fires two pyro cartridges to open the canard surfaces. Gas from the cartridges causes the piston to retract, operating the

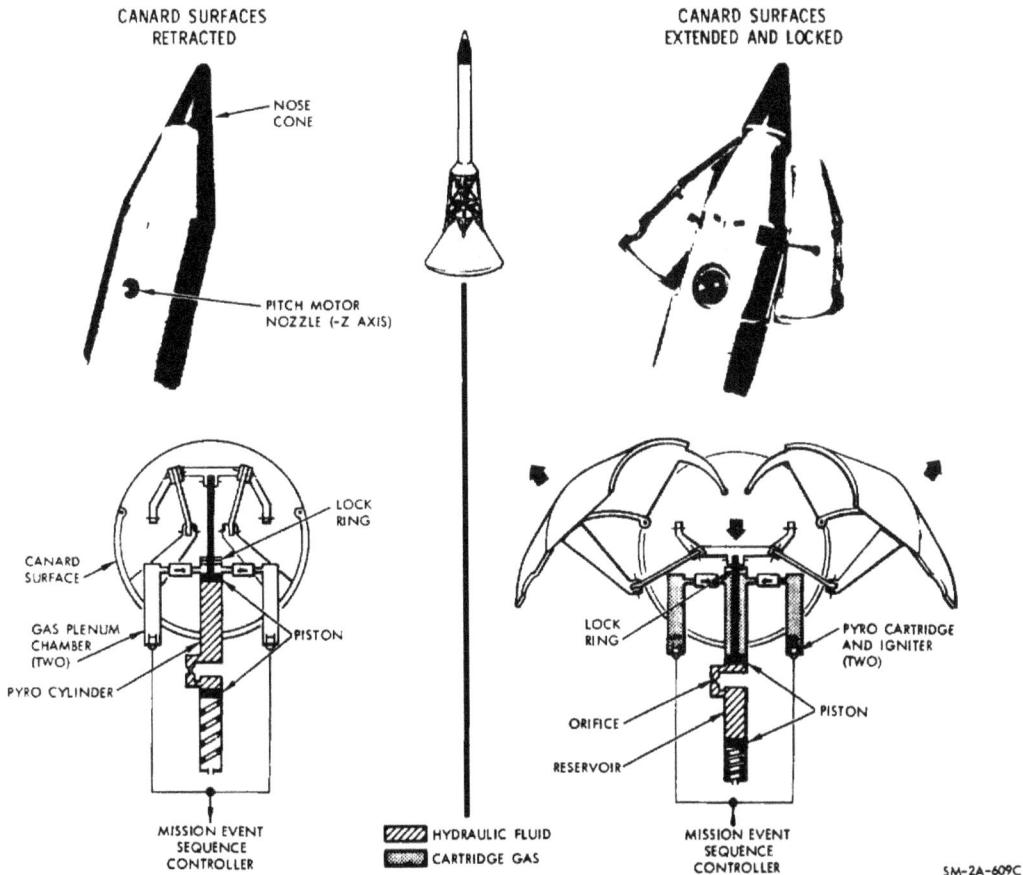

Figure 3-2. Canard Operation

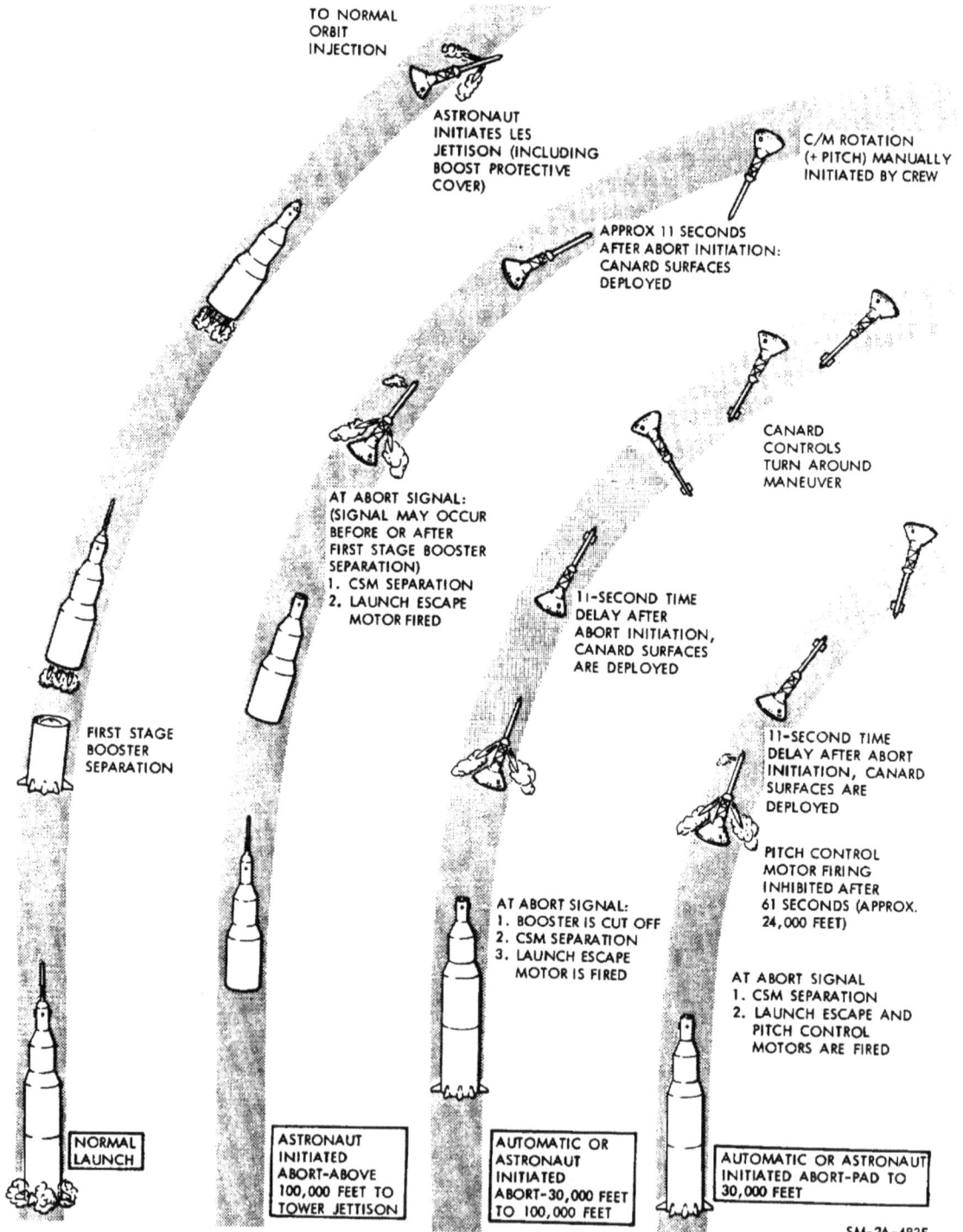

TO NORMAL
ORBIT
INJECTION

ASTRONAUT
INITIATES LES
JETTISON (INCLUDING
BOOST PROTECTIVE
COVER)

C/M ROTATION
(+ PITCH) MANUALLY
INITIATED BY CREW

APPROX 11 SECONDS
AFTER ABORT INITIATION:
CANARD SURFACES
DEPLOYED

CANARD
CONTROLS
TURN AROUND
MANEUVER

AT ABORT SIGNAL:
(SIGNAL MAY OCCUR
BEFORE OR AFTER
FIRST STAGE BOOSTER
SEPARATION)
1. CSM SEPARATION
2. LAUNCH ESCAPE
MOTOR FIRED

11-SECOND TIME
DELAY AFTER
ABORT INITIATION,
CANARD SURFACES
ARE DEPLOYED

11-SECOND TIME
DELAY AFTER ABORT
INITIATION, CANARD
SURFACES ARE
DEPLOYED

FIRST STAGE
BOOSTER
SEPARATION

PITCH CONTROL
MOTOR FIRING
INHIBITED AFTER
61 SECONDS (APPROX.
24,000 FEET)

AT ABORT SIGNAL:
1. BOOSTER IS CUT OFF
2. CSM SEPARATION
3. LAUNCH ESCAPE
MOTOR IS FIRED

AT ABORT SIGNAL
1. CSM SEPARATION
2. LAUNCH ESCAPE AND
PITCH CONTROL
MOTORS ARE FIRED

NORMAL
LAUNCH

ASTRONAUT
INITIATED
ABORT-ABOVE
100,000 FEET TO
TOWER JETTISON

AUTOMATIC OR
ASTRONAUT
INITIATED
ABORT-30,000 FEET
TO 100,000 FEET

AUTOMATIC OR ASTRONAUT
INITIATED ABORT-PAD TO
30,000 FEET

SM-2A-483F

Figure 3-3. Launch Escape and Earth Landing Systems
Functional Diagram (Sheet 1 of 2)

FROM NORMAL ENTRY

AT APPROXIMATELY
24,000 FEET:
1. TOWER SEPARATION
2. TOWER JETTISON MOTOR FIRED
3. BOOST PROTECTIVE COVER
   JETTISONED WITH TOWER
4. APEX COVER JETTISONED 0.4 SECONDS
   AFTER BOOST PROTECTIVE COVER

AT APPROXIMATELY 24,000 FEET
PLUS 0.4 SECONDS
APEX COVER JETTISONED

DROGUE CHUTES DEPLOY
(REEFED) 1.6 SECONDS
AFTER APEX COVER
JETTISONED

DROGUE CHUTES FULLY OPENED
AFTER BEING REEFED FOR 8 SECONDS

APEX COVER
JETTISONED 0.4
SECONDS AFTER
LES TOWER
JETTISON

DROGUE CHUTES DEPLOYED
(REEFED) 2 SECONDS AFTER
LES TOWER JETTISON

DROGUE CHUTES
RELEASED AND
PILOT CHUTE
MORTARS FIRED
TWELVE SECONDS
AFTER DROGUE
CHUTE
DEPLOYMENT
OR AT
APPROXIMATELY
10,000 FEET

3-SECOND TIME DELAY AFTER
CANARD DEPLOYMENT:
1. TOWER SEPARATION
2. TOWER JETTISON MOTOR
   FIRED
3. BOOST PROTECTIVE COVER
   JETTISONED WITH TOWER

MAIN CHUTES EXTRACTED
& DEPLOYED TO A REEFED
CONDITION

MAIN CHUTES FULLY
OPENED AFTER BEING
REEFED FOR 8 SECONDS

NOTE: SATURN V BOOSTER
SHOWN IN DIAGRAM.

MAIN CHUTES RELEASED
AFTER TOUCHDOWN

SM-2A-473G

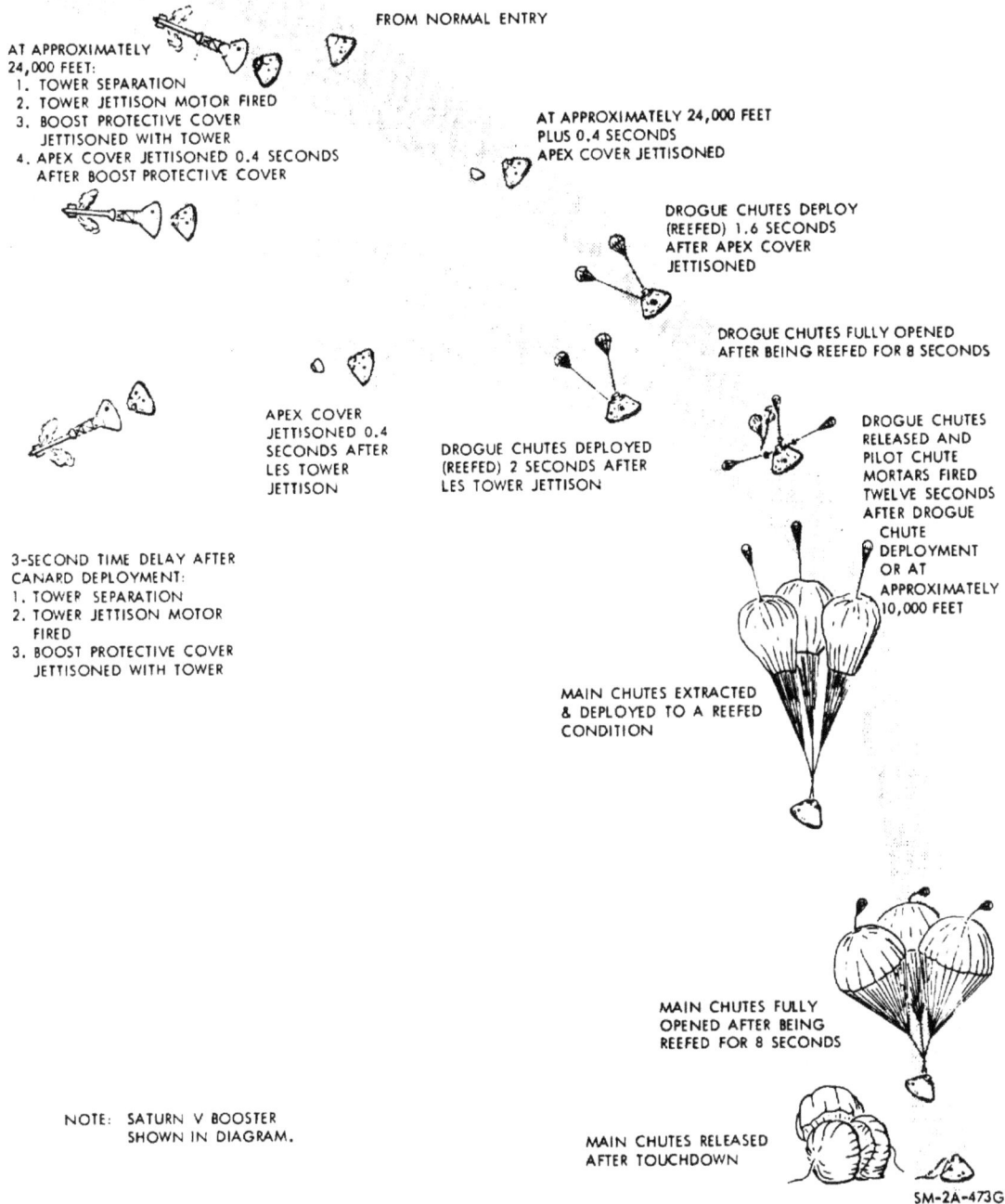

Figure 3-3. Launch Escape and Earth Landing Systems
Functional Diagram (Sheet 2 of 2)

opening mechanism. The cylinder is filled with hydraulic fluid downstream of the piston and, as the piston retracts, the fluid is forced out through an orifice into a reservoir. Metering the fluid through the orifice controls the speed of the canard operation. When fully open, the two canard surfaces are locked in place by gas pressure in the cylinder, a lock ring on the piston shaft, and an overcenter linkage. The induced aerodynamic forces acting on the canard surfaces will orient the C/M to a blunt-end-forward trajectory before launch escape assembly jettison and ELS activation. (See figure 3-3.)

3-11. EMERGENCY DETECTION SYSTEM.

3-12. The emergency detection system (EDS) is designed to detect and display emergency conditions of the launch vehicle to the astronaut. The EDS also provides automatic abort initiation, under certain conditions, between lift-off and 90 seconds from launch. The display circuitry and automatic abort capabilities are enabled at lift-off. A lockout system is provided to prevent enabling the automatic abort circuitry prior to lift-off. The astronaut may initiate an abort manually at any time with the commanders translation control.

3-13. AUTOMATIC ABORT.

3-14. The emergency detection system (EDS) will initiate an automatic abort signal after launch by sensing excessive vehicle rates or two engines out. The abort signal will cause booster cutoff, event timer reset, and launch escape system activation.

3-15. MANUAL ABORT.

3-16. A manual abort (figure 3-4) can be initiated prior to, or during launch by manual CCW rotation of the commanders translation control. The launch escape system can be utilized until approximately normal tower jettison. During a normal mission, the LES tower is jettisoned shortly after second booster stage ignition (280,000 feet with uprated Saturn I or 320,000 feet with Saturn V), and any abort thereafter is accomplished by utilizing the SPS engine in the service module. An SPS abort must be manually initiated. Upon abort initiation, the booster automatically separates from the spacecraft, S/M-RCS engines fire to accomplish the ullage maneuver, and the SPS engine ignites to thrust the S/C away from the booster.

3-17. ABORT REQUEST INDICATOR LIGHT AND EVENT TIMER.

3-18. The ABORT request indicator light is illuminated by ground control, or the range safety officer, using GSE or a radio command through the up-data link. When illuminated, the light indicates an abort request and serves to alert the crew of an emergency situation.

3-19. Initiation of an abort (automatic or manual) will automatically reset the event timer to provide a time reference for manual operations.

3-20. EARTH LANDING SYSTEM.

3-21. The purpose of the earth landing system (ELS) is to provide a safe landing for the astronauts and the command module following an abort, or a normal entry from an earth orbital or lunar landing mission. (See figure 3-3.) Included as a part of the ELS are several recovery aids which are activated after impact on either land or water. The ELS operation is automatically timed and activated by sequence controllers. There are, however, backup components and manual override controls to ensure system reliability and to provide astronaut control.

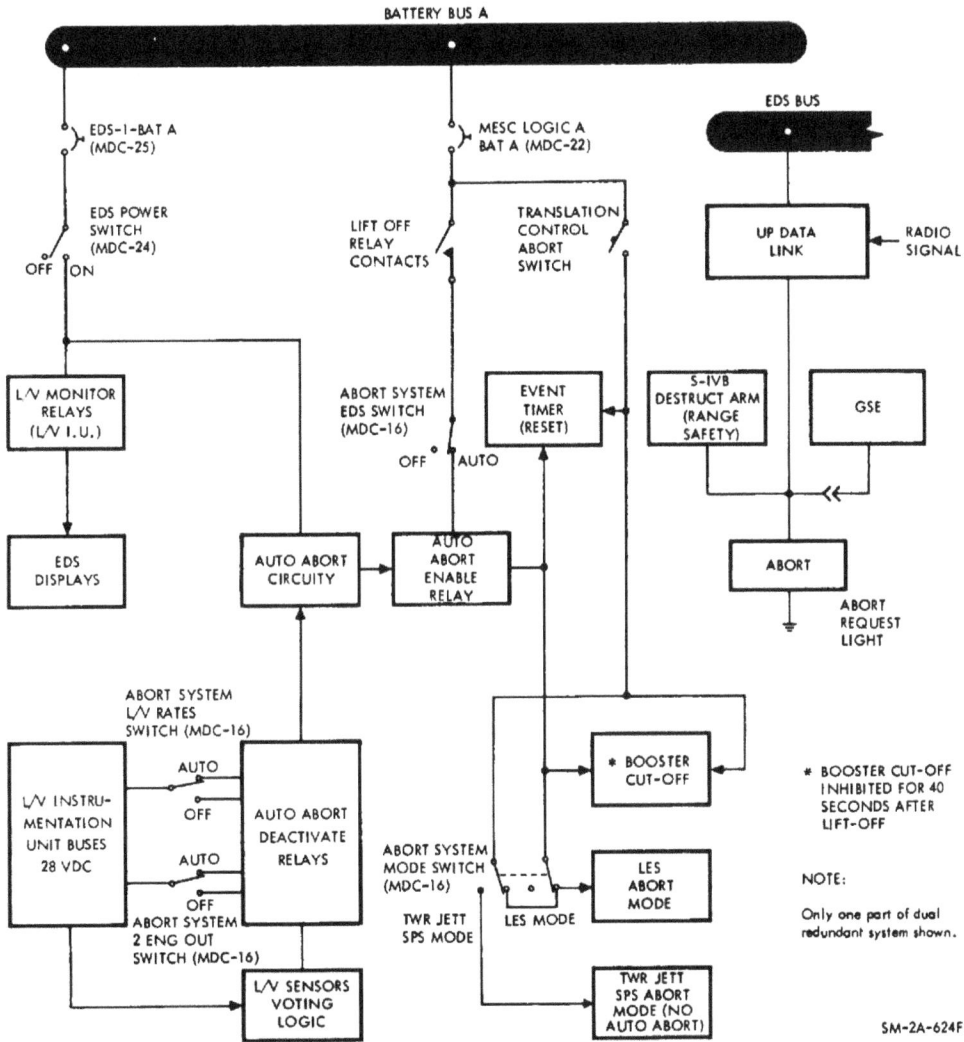

Figure 3-4. Emergency Detection System Automatic and Manual
Abort Block Diagram

3-22. With the exception of the controls and sequence controllers in the C/M crew compartment, ELS components are located in the forward compartment of the C/M as shown in figure 3-5. The ELS consists of the forward heat shield ejection subsystem, the sequence controllers, recovery aids, and the parachute subsystem. The parachute subsystem, is comprised of two fist ribbon nylon drogue parachutes, 13.7 feet in diameter; three ring slot nylon pilot parachutes, 7.2 feet in diameter; three ring sail nylon main parachutes, 83.5 feet in diameter; deployment bags, harness, mortars, and the necessary hardware for attachment to the C/M.

3-23. ELS OPERATION.

3-24. The C/M-ELS begins operation upon descending to approximately 24,000 feet +0.4 second, or in the event of an abort, 0.4 second after launch escape assembly jettison. (See figure 3-3.) The apex cover (forward heat shield) is jettisoned by four gas-pressure thrusters. This function is imperative, as the forward heat shield covers and protects the ELS parachutes up to this time. At 1.6 seconds later, the drogue mortar pyrotechnic cartridges are fired to deploy two drogue parachutes in a reefed condition. After 8 seconds, the reefing lines are severed by reefing line cutters and the drogue parachutes are fully opened. These stabilize the C/M in a blunt-end-forward attitude and provide deceleration. At approximately 10,000 feet, drogue parachutes are released, and the three pilot parachute mortars are fired. This action ejects the pilot parachutes which extract and deploy the three main parachutes. To preclude the possibility of parachute damage or failure due to the descent velocity, the main parachutes open to a reefed condition for 8 seconds to further decelerate the C/M. The three parachutes are then fully opened (disreefed) to lower the C/M at a predetermined descent rate. At 27-1/2-degree hang angle of the C/M is achieved by the main parachute attachment points. In the event one main parachute fails to open, any two parachutes will safely carry out the prescribed function.

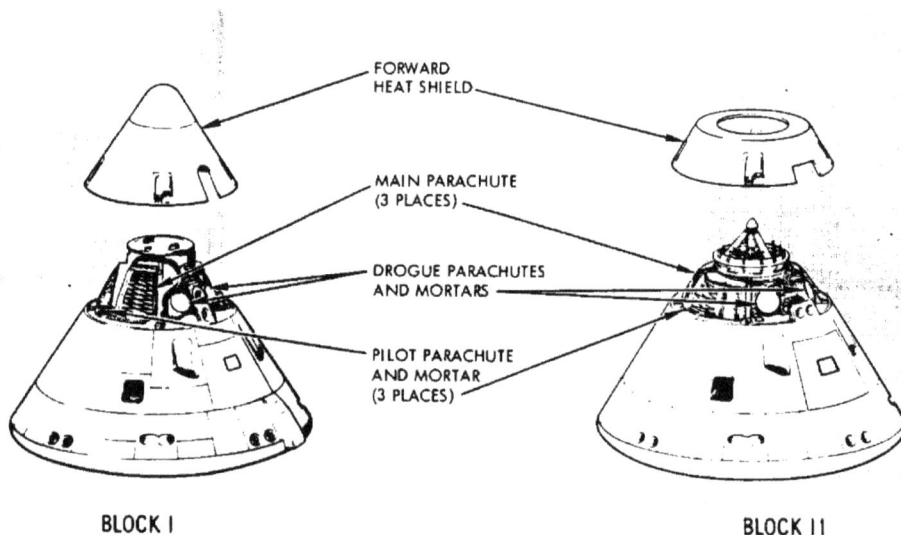

SM-2A-482E

Figure 3-5. Earth Landing System

The main parachutes are disconnected following impact. The recovery aids consists of an uprighting system, swimmers umbilical, sea (dye) marker, a flashing beacon light, a VHF recovery beacon transmitter, a VHF transceiver, and an H-F transceiver. A recovery loop is also provided on the C/M to facilitate lifting. If the command module enters the water and stabilizes in a stable II (inverted) condition, the uprighting system is activated (manually), inflating three air bags causing the command module to assume a stable I (upright) condition. Each bag has a separate switch for controlling inflation. The sea (dye) marker and swimmer's umbilical are deployed automatically when the HF recovery antenna is deployed (manually initiated by crew). The marker is tethered to the C/M forward compartment deck and will last approximately 12 hours. The swimmer's umbilical provides the electrical connection for communication between the crew in the C/M and the recovery personnel in the water.

3-25. ENVIRONMENTAL CONTROL SYSTEM.

3-26. The basic purpose of the environmental control system (ECS) is to provide a controlled environment for three astronauts in the Apollo spacecraft during missions of up to 14 days duration. This environment consists of a pressurized suit circuit for use during normal or emergency conditions, and a pressurized shirtsleeve cabin atmosphere used only when normal conditions exist. Metabolically, the system is responsible for supplying oxygen and hot and cold potable water; as well as removing carbon dioxide, odors, water production, and heat output. The ECS also disperses electronic equipment heat loads.

3-27. ECS OPERATION.

3-28. The ECS is so designed that a minimum amount of crew time is required for normal system operation. Electrical and mechanical override and backup capabilities exist throughout the system to maintain the required reliability level. Oxygen and potable water are supplied to the ECS by components of the electrical power system (EPS). (See figure 3-6.) The oxygen originates in the cryogenic storage tanks and the potable water is a by-product of the fuel cell modules. All are located in the S/M. Waste water is collected from moisture that condenses within the pressure suit circuit, and is stored for ECS utilization. Block I S/C incorporates additional water tanks in the S/M. This supply permits the accomplishment of maximum duration earth orbital missions.

3-29. In maintaining the pressure suit and cabin shirtsleeve environments, the ECS continually conditions the atmospheres of both. This is accomplished by automatically controlling the supply of oxygen; regulating the flow, pressure, temperature, and humidity; and removing unwanted items such as carbon dioxide, odors, heat, and moisture. The pressure suit and cabin gases are processed for re-use by being routed through the suit circuit debris trap, CO_2-odor absorber filters, and heat exchanger. Additional components installed on Block II S/C permit the CSM ECS to pressurize the LM.

3-30. A continuous circulating mixture of water-glycol provides the ECS with a heat transport fluid loop. This flow is used to cool the cabin atmosphere, pressure suit circuit, electronic equipment, and a portion of the potable water. It also serves as a heat source for the cabin, when required. All unwanted heat absorbed by the water-glycol is transported to the ECS space radiators located on the surface of the S/M. (These radiators are not to be confused with the EPS radiators also located on this surface.) Should cooling by radiation be inadequate, supplemental cooling takes place within the water-glycol evaporator where heat is rejected by the evaporation of waste water. In addition to this primary coolant loop, Block II S/C are provided with a completely independent, secondary coolant loop. The redundant loop routes water-glycol to certain critical components that are absolutely necessary to complete a safe return to earth.

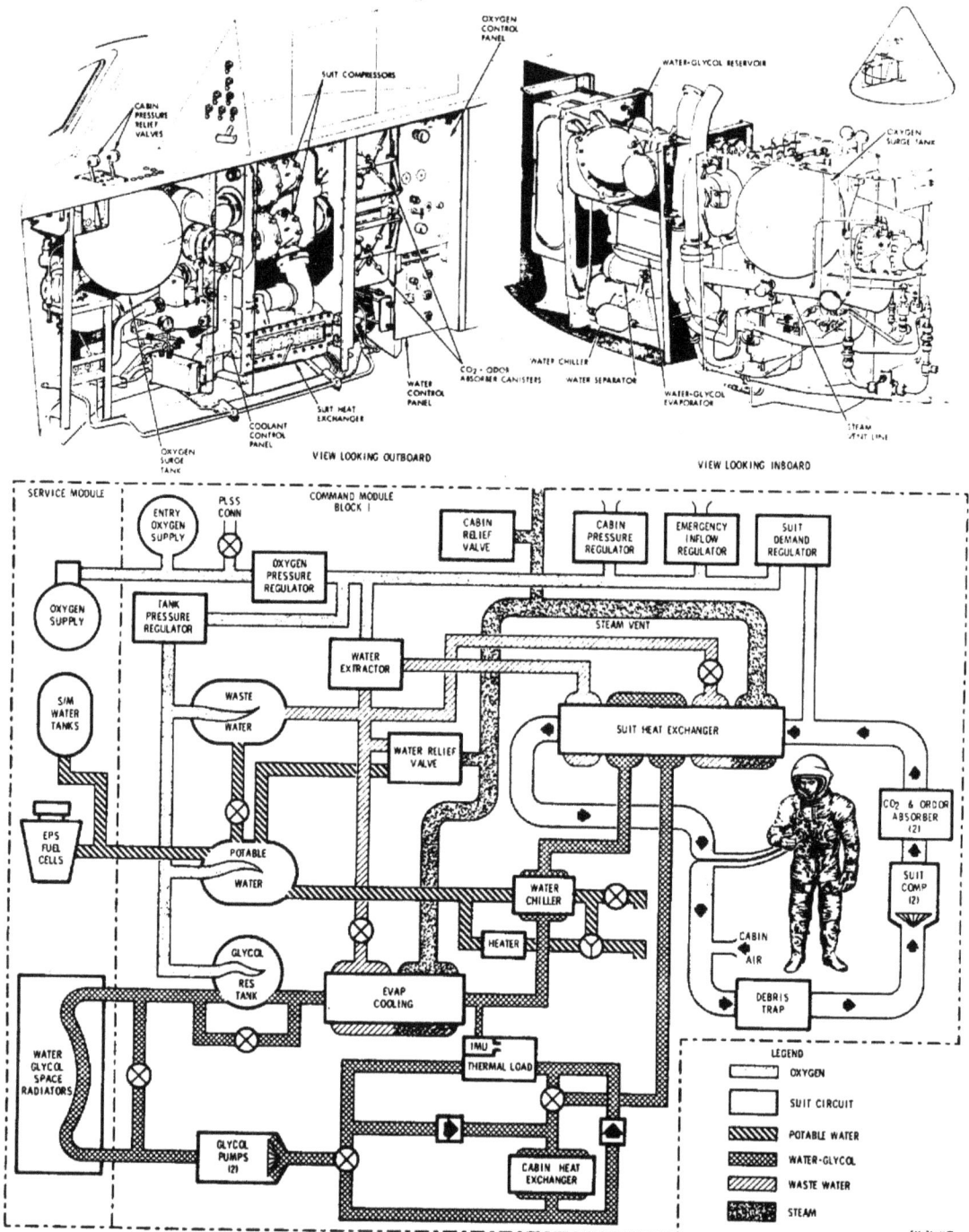

Figure 3-6. Environmental Control System Simplified Flow Diagram

3-31. The water supply subsystem is concerned with the storage and distribution of potable water produced by the fuel cell modules, and waste water recovered from the suit heat exchanger. These supplies are used by the ECS to furnish hot and cold potable water for crew consumption, and waste water for evaporative cooling by the water-glycol evaporator and suit heat exchanger (Block I only).

3-32. ECS ENTRY PROVISION. CSM separation prior to entry removes the capability of cooling the water-glycol by space radiation, and cuts off the source of potable water and oxygen. Provision is therefore made to enable the ECS to carry out its functions following separation. Before separation occurs, the C/M cabin is cold-soaked using a flow of cold water-glycol. This provides a heat sink for the aerodynamically developed heat that penetrates the C/M during entry. After separation, the water-glycol and the suit circuit are cooled exclusively by water evaporation. A tank in the C/M supplies all the oxygen required during entry and descent. Upon landing, a postlanding ventilation system is activated. This consists of two valves and a fan which will circulate ambient air through the C/M. The postlanding ventilation system is activated by the crew.

3-33. ELECTRICAL POWER SYSTEM.

3-34. The primary purpose of the electrical power system (EPS) is to provide the electrical energy sources, power generation and controls, power conversion and conditioning, and power distribution to a-c and d-c electrical buses to meet the requirements of the various spacecraft systems during the mission flight and postlanding phases. This is shown in figures 3-7 and 3-9 and described in paragraph 3-43. For ground checkout, all d-c electrical power will be supplied by ground support equipment prior to activation of the fuel cells. During this same period, a-c electrical power will be supplied by the S/C inverters for some checkout functions and by ground support equipment for other checkout functions. The secondary purpose of the EPS is to furnish the environmental control system with potable water required by the three astronauts during the mission. This water is obtained as a by-product of the three fuel cell powerplants. Paragraph 3-43 provides a list of the various power sources and the power-consuming devices connected to each bus.

3-35. D-C POWER SUPPLY.

3-36. Two d-c power sources provide the S/C with 28-volt d-c power. The first source consists of three Bacon-type, hydrox (hydrogen-oxygen) fuel cell powerplants. The second source is obtained from silver oxide-zinc-type storage batteries. The fuel cell powerplants are connected in parallel and used throughout the mission until command-service module (CSM) separation. Any two of the three fuel cells will provide sufficient power for normal mission loads. In the event two fuel cells fail, the third is capable of furnishing emergency power; however, this is contingent upon removing all nonessential loads from the bus, in addition to supplying battery power to the bus at peak loads above the capacity of the one operating fuel cell powerplant. The d-c power system is controlled, regulated, and protected by appropriate switching circuits, undervoltage detection, and circuit breakers. Two nonrechargeable batteries (Block I only), the fuel cell powerplants, the cryogenic storage system for the hydrogen and oxygen, and the glycol coolant space radiators are located in the service module. (See figure 3-8.)

3-37. CRYOGENIC GAS STORAGE SYSTEM. The cryogenic gas storage system supplies the hydrogen and oxygen consumed by the three fuel cell powerplants. In addition, the oxygen used by the ECS is also supplied from this source. The hydrogen and oxygen subsystems are very similar, each consisting of storage tanks, associated valves, pressure

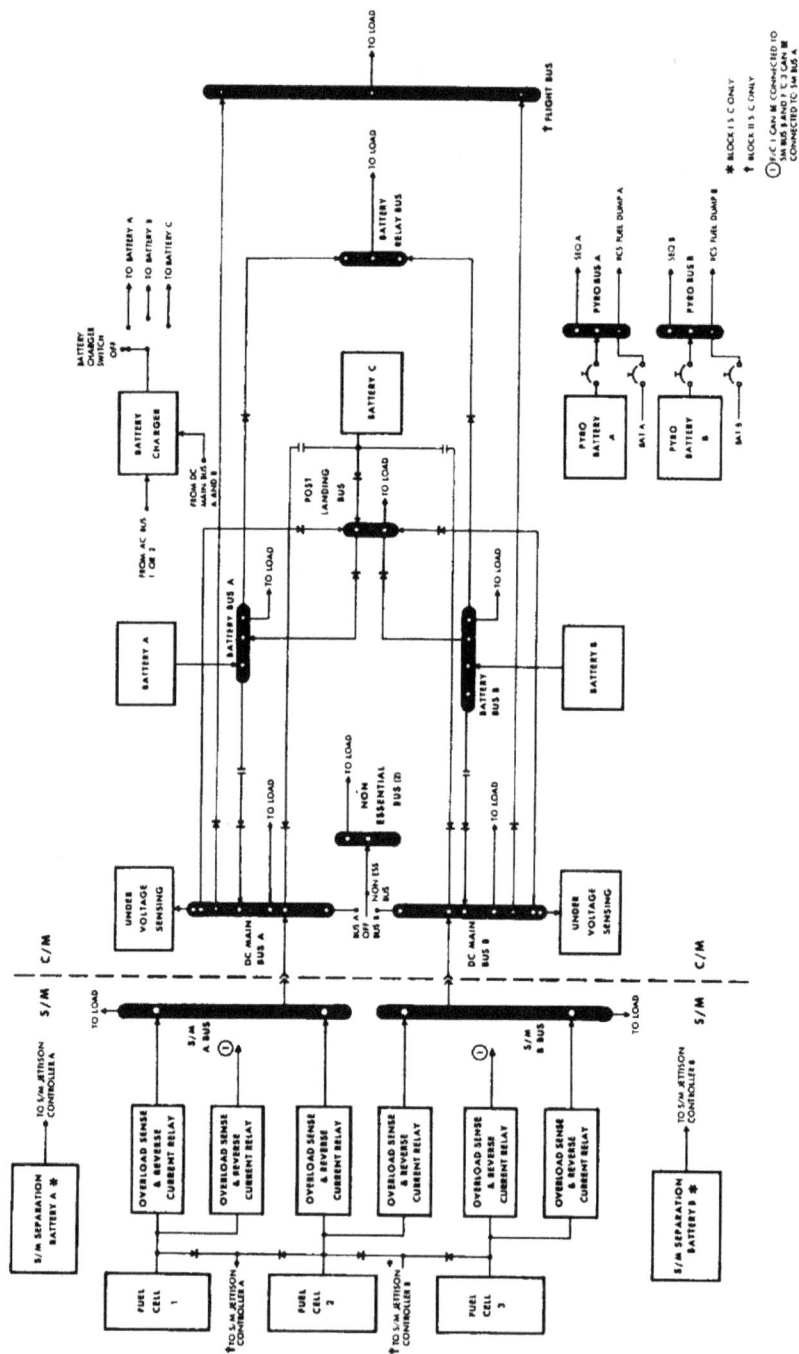

Figure 3-7. Electrical Power System – D-C Power Distribution Diagram

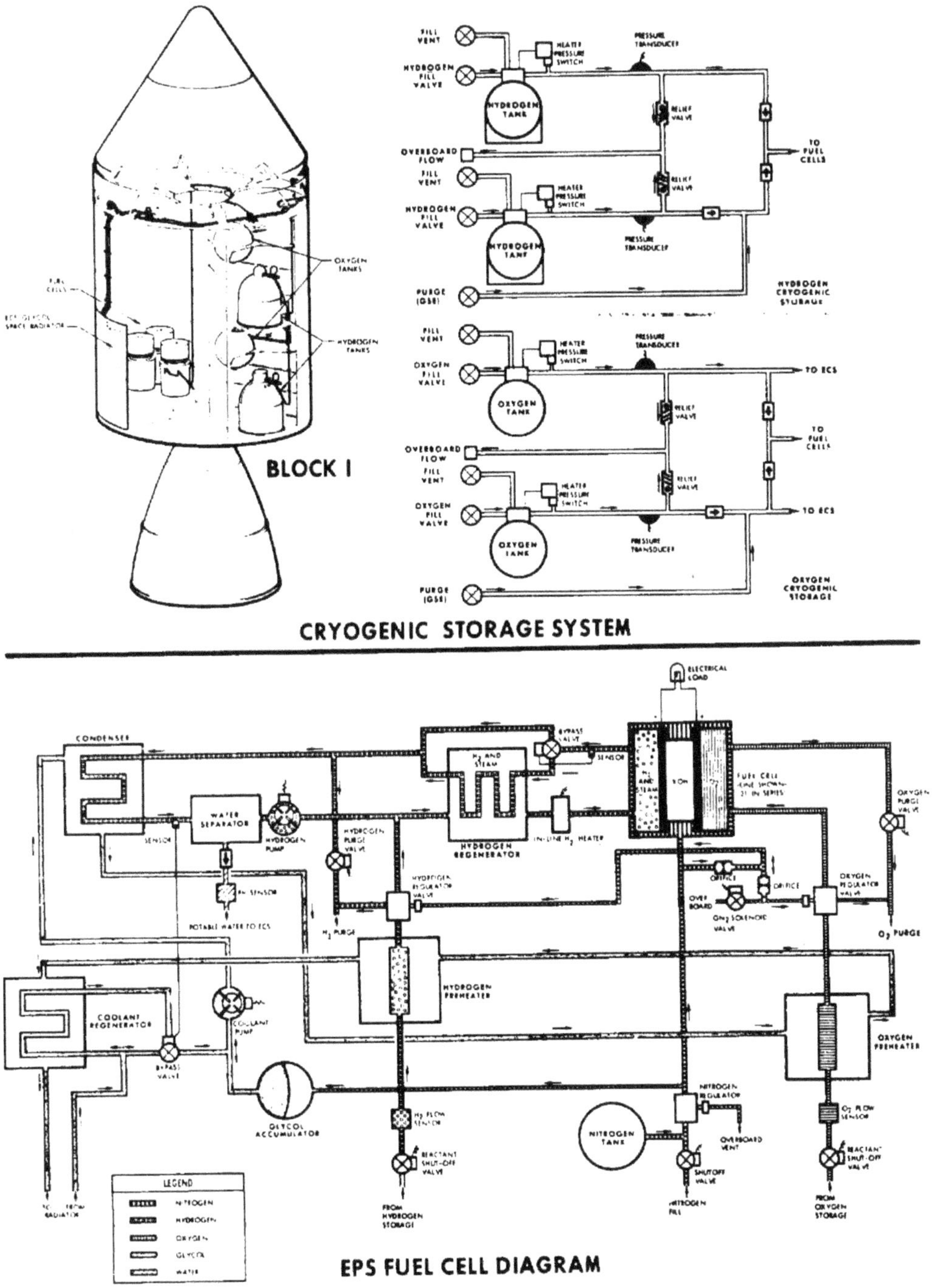

BLOCK I

CRYOGENIC STORAGE SYSTEM

EPS FUEL CELL DIAGRAM

Figure 3-8. Electrical Power System — Cryogenics Storage
and Fuel Cell Functional Diagram

switches, motor switches, lines, and other plumbing components. (See figure 3-8.) The hydrogen and oxygen are stored in a cryogenic state. However, by the time both reactants reach the fuel cells, they have warmed considerably and are in a gaseous state.

3-38. FUEL CELL POWERPLANTS. Each of the three fuel cell powerplants consist of 31 single cells connected in series. Each cell consists of a hydrogen compartment, an electrolyte compartment, an oxygen compartment, and two electrodes. (See figure 3-8.) The electrolyte is composed of potassium hydroxide (78 percent) and water (22 percent) and remains constant in cell reaction by simply providing an ionic conduction path between the electrodes. The hydrogen electrode is composed of nickel, while the oxygen electrode is composed of nickel and nickel oxide. This electrode structure also remains constant throughout fuel cell operation. The consumable reactants of hydrogen and oxygen are supplied to the cell under regulated pressure, using nitrogen pressure as a reference as well as for pressurizing the powerplants. By chemical reaction, electricity, water, and heat are produced, with the reactants being consumed in proportion to the electrical load. The by-products, water and heat, are utilized to maintain the supply of potable water and to keep the electrolyte at the proper operating temperature. The 31 fuel cells and the required pumps, valves, regulators, and other components are housed in a container.

3-39. BATTERIES. Three batteries, located in the lower equipment bay of the C/M, can be selected and switched to a variety of buses and circuits. The circuitry which ignites the S/C pyrotechnic devices is generally completely independent and isolated from the remainder of the d-c system, and receives power from two pyrotechnic batteries. There are two nonrechargeable batteries in the S/M (Block I only) whose sole function is to furnish power to the service module jettison controllers. In Block II, S/C power to the S/M jettison controllers is furnished by the fuel cell powerplants. This will sustain the firing of those S/M reaction control engines that provide S/M retrograde following C/M—S/M separation.

3-40. BATTERY CHARGER. A battery charger, located in the lower equipment bay of the C/M, is utilized to assure that the entry batteries are fully charged before entry begins.

3-41. A-C POWER SUPPLY.

3-42. Three solid-state inverters, located in the lower equipment bay of the C/M, are the source of the 115/200-volt 400-cycle 3-phase a-c power used in the S/C. These inverters operate from the two 28-volt d-c main buses, and supply power to two 400-cycle a-c buses. (See figure 3-9.) The a-c electrical power system is complete with adequate switching arrangements, overvoltage and overload sensing circuits, as well as circuit breakers for protection of the inverters. Under normal conditions, one inverter has the capability of supplying all S/C primary 400-cycle a-c electrical power needs. The other two inverters act as redundant standby units. In the event of inverter failure, input and load are manually switched to another inverter. Although the inverters cannot be paralleled on a single bus due to circuitry provisions, two inverters can operate simultaneously if each supplies a separate a-c bus.

3-43. SPACECRAFT POWER SOURCES AND POWER CONSUMING DEVICES.

3-44. The following list contains Block I and Block II spacecraft power sources and power-consuming devices.

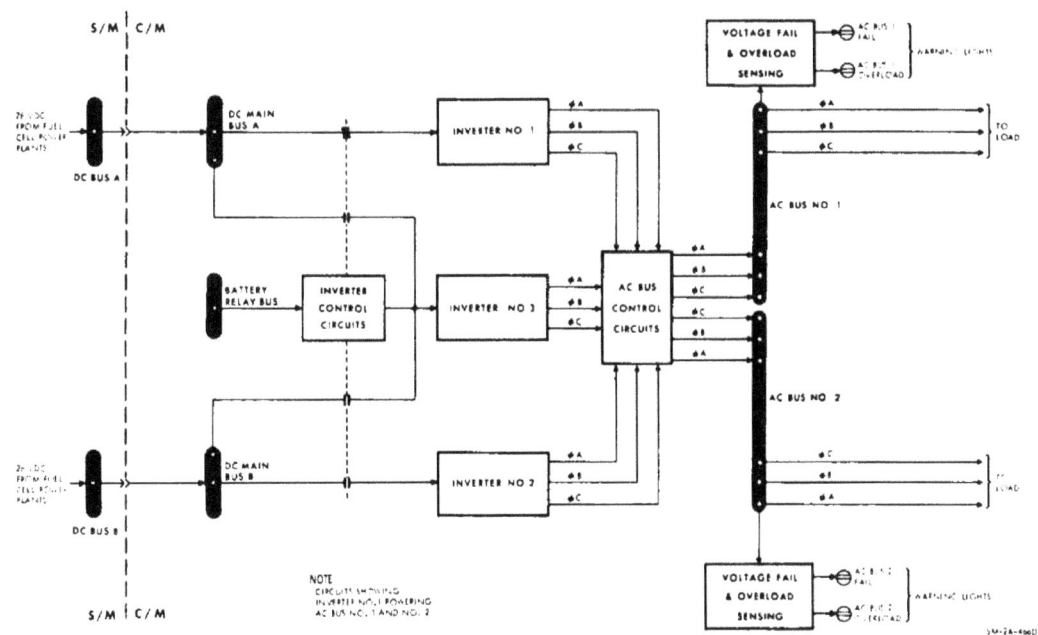

Figure 3-9. Electrical Power System — A-C Power Distribution Diagram

Block I	Block II
Command Module D-C Main Bus A (Powered by fuel cells 1, 2, and 3, and backed up by batteries A, and C, when necessary)	
Environmental control system	Environmental control system
Pressure and temperature transducers	Pressure and temperature transducers
Water separator No. 1	Emergency loop temperature transducers
S/M water tank control	Water accumulator
Steam duct heater No. 1	Steam duct heater No. 1 and water-glycol temperature control
Flight and postlanding bus	Compressor inverter
Pyro interrupter switch	
Oxygen and hydrogen purge - fuel cell power plants No. 1, 2, and 3	Flight and postlanding bus
Inverters No. 1 and No. 3	Pyro interrupter switch
Battery charger	Oxygen and hydrogen purge - fuel cell powerplants No. 1, 2, and 3
Nonessential bus switch	Inverters No. 1 and No. 3

Block I	Block II
Interior floodlighting	Battery charger
D-C sensing unit and voltmeter switch	Nonessential bus switch
Stabilization and control system	Interior floodlighting
Direct control	D-C sensing unit and voltmeter switch
Pitch	Stabilization and control system
Roll-channel A&C	Direct control
Roll-channel B&D	Pitch
Yaw	Roll-channel A&C
Group 1	Roll-channel B&D
Group 2	Yaw
	Logic
Potable water heater	Potable water heater
Caution and warning detection unit	Caution and warning detection unit
Event timer	Event timer
Central timing equipment	Central timing equipment
Reaction control system	Reaction control system
Propellant isolation	Propellant isolation
RCS transfer	RCS transfer
RCS heaters	RCS heaters
Essential instrumentation	Essential instrumentation
Cryogenic oxygen and hydrogen tank heaters	Cryogenic oxygen and hydrogen tank heaters
Service propulsion system	Service propulsion system
Gauging	Gauging
Helium shutoff valve	Helium shutoff valve

Block I	Block II
Guidance and navigation system	Guidance and navigation system
Inertial measurement unit-coupling display unit	Inertial measurement unit-coupling display unit
Inertial measurement unit heaters	Inertial measurement unit heaters
Optics	Optics
Computer	Computer
Space suit communications and biomed instrumentation	Entry monitor display
Crew couch attenuation	Flight bus
	Crew couch attenuation
	Rendezvous radar transponder
	Docking lights

Command Module D-C Main Bus B (Powered by fuel cells 1, 2, and 3, and backed up by batteries B, and C, when necessary)

Block I	Block II
Environmental control system	Environmental control system
Pressure and temperature transducers	Pressure and temperature transducers
Water separator No. 2	Emergency loop temperature transducers
S/M water tank control	Water accumulator
Steam duct heater No. 2	Steam duct heater No. 2 and water-glycol temperature control
Flight and postlanding bus	Flight and postlanding bus
Pyro interrupter switch	Pyro interrupter switch
Oxygen and hydrogen purge - fuel cell powerplants No. 1, 2, and 3	Oxygen and hydrogen purge - fuel cell powerplants No. 1, 2, and 3
Inverters No. 2 and No. 3	Inverters No. 2 and No. 3
Battery charger	Battery charger
Nonessential bus switch	Nonessential bus switch
Interior floodlighting	Interior floodlighting
D-C sensing unit and voltmeter switch	

Block I	Block II
Stabilization and control system	D-C sensing unit and voltmeter switch
Direct control	Stabilization and control system
Pitch	Direct control
Roll-channel A&C	Pitch
Roll-channel B&D	Roll-channel A&C
Yaw	Roll-channel B&D
Group 1	Yaw
Group 2	Logic
Potable water heater	Potable water heater
Caution and warning detection unit	Caution and warning detection unit
Event timer	Event timer
Central timing equipment	Central timing equipment
Reaction control system	Reaction control system
Propellant isolation	Propellant isolation
RCS transfer	RCS transfer
RCS heaters	RCS heaters
Essential instrumentation	Essential instrumentation
Cryogenic oxygen and hydrogen tank heaters	Cryogenic oxygen and hydrogen tank heaters
Service propulsion system	Service propulsion system
Gauging	Gauging
Helium shutoff valve	Helium shutoff valve
Guidance and navigation system	Guidance and navigation system
Inertial measurement unit-coupling display unit	Inertial measurement unit-coupling display unit
Inertial measurement unit heaters	Inertial measurement unit heaters

Block I	Block II
Optics	Optics
Computer	Computer
Space suit communications and biomed instrumentation	Entry monitor display
Crew couch attenuation	Crew couch attenuation
	Rendezvous radar transponder
	Docking lights
	LM power switch
Command Module A-C Bus No. 1 (Powered by inverter No. 1, 2, or 3)	
Reactant pump - fuel cell powerplants No. 1, 2, and 3	Reactant pump - fuel cell powerplants No. 1, 2, and 3
Battery charger	Battery charger
Stabilization and control - group 1 and group 2	Stabilization and control - group 1 and group 2
Guidance and navigation system	Guidance and navigation system
Cryogenic fuel quantity amplifier	Cryogenic fuel quantity amplifier
Telecommunications	Telecommunications
Environmental control system	Environmental control system
Glycol pumps	Glycol pumps
Suit compressors	Suit compressors
Cabin air fans, water-glycol temperature control, suit temperature control, and waste management blower	Cabin air fans, water-glycol temperature control, suit temperature control, and waste management blower
Space radiator isolation valves	Water-glycol emergency loop
Interior lighting	Space radiator isolation valves
SPS gauging	Interior lighting
A-C sensing unit and voltmeter switch	Exterior lighting
Cryogenic oxygen and hydrogen tank fan motors (system 1)	SPS gauging

Block I	Block II
Gas analyzer	A-C sensing unit and voltmeter switch
	Cryogenic oxygen and hydrogen tank gan motors (system 1)

Command Module A-C Bus No. 2 (Powered by inverter No. 1, 2, or 3)	
Reactant pump - fuel cell powerplants No. 1, 2, and 3	Reactant pump - fuel cell powerplants No. 1, 2, and 3
Battery charger	Battery charger
Stabilization and control - group 1 and group 2	Stabilization and control - group 1 and group 2
Guidance and navigation system	Guidance and navigation system
Cryogenic fuel quantity amplifier	Cryogenic fuel quantity amplifier
Telecommunications	Telecommunications
SPS gauging	SPS gauging
Environmental control system	Environmental control system
Glycol pumps	Glycol pumps
Space radiator isolation valves	Space radiator isolation valves
Cabin air fans, cabin temperature control and water-glycol temperature control	Cabin air fans, cabin temperature control and water-glycol temperature control
Suit compressors	Water-glycol emergency loop
Interior lighting	Suit compressors
A-C sensing unit and voltmeter switch	EVT oxygen valve
Cryogenic oxygen and hydrogen tank fan motors (system 2)	Interior lighting
	Exterior lighting
	A-C sensing unit and voltmeter switch
	Cryogenic oxygen and hydrogen tank fan motors (system 2)

Block I	Block II
Command Module Battery Bus A (Powered by entry battery A)	
Flight and postlanding bus	Flight and postlanding bus
ELS sequencer A logic and SECS arm	ELS sequencer A logic and SECS arm
Battery charger and battery bus A tie switch	Battery charger and battery bus A tie switch
Arm mission sequencer logic bus and EDS abort enable switch	Battery relay bus
Logic mission sequencer A and voltmeter switch	EDS - bus No. 1
	D-C main bus A
Battery relay bus	D-C sensing unit and voltmeter switch
EDS - bus No. 1	Main gimbal control - yaw, pitch
D-C main bus A	Uprighting system - compressor No. 1
D-C sensing unit and voltmeter switch	Flotation bag control
Main gimbal control - yaw, pitch	
Uprighting system - compressor No. 1	
Command Module Battery Bus B (Powered by entry battery B)	
Flight and postlanding bus	Flight and postlanding bus
ELS sequencer B logic and SECS arm	ELS sequencer B logic and SECS arm
Battery charger and battery bus B tie switch	Battery charger and battery bus B tie switch
Arm mission sequencer logic bus and EDS abort enable switch	Battery relay bus
Logic mission sequencer B and voltmeter switch	ECS - bus No. 3
	D-C main bus B
Battery relay bus	D-C sensing unit and voltmeter switch
EDS - bus No. 3	Auxiliary gimbal control - yaw, pitch
D-C main bus B	Uprighting system - compressor No. 2
D-C sensing unit and voltmeter switch	Flotation bag control

Block I	Block II
Auxiliary gimbal control - yaw, pitch	
Uprighting system - compressor No. 2	

Command Module Flight and Postlanding Bus (Powered by entry battery C, d-c main buses A and B, and battery buses A and B)

Block I	Block II
VHF recovery beacon	D-C main bus A
D-C main bus A	D-C main bus B
D-C main bus B	Microphone amplifiers — NAV, CMDR, ENGR
Audio center (engineer)	Floodlights
Audio center transmitter key relay	ECS postlanding ventilation system
VHF/AM transmitter receiver	Flotation bag No. 3
H-F transceiver	EDS - bus No. 2
Audio center (CMDR)	
Up-data link	
VHF/FM transmitter	
S-band power amplifier	
Unified S-band power relay	
Signal conditioning equipment (Block I)	
TV camera	
C-band transponder	
Data storage equipment	
Premodulation processor	
Audio center (NAV)	
Microphone amplifiers — NAV, CMDR, ENGR	
ECS postlanding ventilation system	
Flotation bag control	

Block I	Block II
Command Module Battery Relay Bus (Powered by entry batteries A and B)	
Control circuits - inverters No. 1, 2, and 3	Control circuits - inverters No. 1, 2, and 3
A-C buses No. 1 and No. 2 over-undervoltage and overload sensing	A-C buses No. 1 and No. 2 over-undervoltage and overload sensing
Reactant shutoff valves - fuel cell powerplants No. 1, 2, and 3	Reactant shutoff valves - fuel cell powerplants No. 1, 2, and 3
D-C main buses A and B undervoltage sensing unit	D-C main buses A and B undervoltage sensing unit
D-C main buses A and B select switch, fuel cell powerplants No. 1, 2, and 3 and indicators	D-C main buses A and B select switch, fuel cell powerplants No. 1, 2, and 3 and indicators
	Fuel cell powerplants No. 1, 2, and 3 radiator valves
Command Module Nonessential Buses (Powered by d-c main bus A or B)	
Nonessential instrumentation	Nonessential instrumentation
NASA scientific instrumentation	NASA scientific instrumentation
Flight qualification recorder	Special equipment bays No. 1 and No. 2
Special equipment bays No. 1 and No. 2 (S/C 012 and S/C 014)	Special equipment hatch
Special equipment hatch (S/C 012 and S/C 014)	
Command Module MESC Pyro Bus A (Powered by pyro battery A)	
Sequencer A	Sequencer A
RCS fuel dump and voltmeter switch	HF orbital antenna deploy
LES, ELS, and RCS pressure initiators	RCS fuel dump and voltmeter switch
	LES, ELS, and RCS pressure initiators

Block I	Block II
Command Module MESC Pyro Bus B (Powered by pyro battery B)	
Sequencer B	Sequencer B
RCS fuel dump and voltmeter switch	HF recovery antenna deploy
LES, ELS, and RCS pressure initiators	RCS fuel dump and voltmeter switch
	LES, ELS, and RCS pressure initiators
Command Module Entry Battery C	
D-C main bus A	D-C main bus A
D-C main bus B	D-C main bus B
Flight and postlanding bus	Flight and postlanding bus
EDS - bus No. 2	Voltmeter switch
Voltmeter switch	
Service Module D-C Bus A (Powered by fuel cells 1, 2, and 3)	
SPS primary gimbal motors	SPS primary gimbal motors
Overload or reverse current sensing - fuel cell powerplants No. 1, 2, and 3	Overload or reverse current sensing - fuel cell powerplants No. 1, 2, and 3
D-C main bus A	S/M jettison controller A
	D-C main bus A
Service Module D-C Bus B (Powered by fuel cell 1, 2, or 3)	
SPS auxiliary gimbal motors	SPS auxiliary gimbal motors
Overload or reverse current sensing - fuel cell powerplants No. 1, 2, and 3	Overload or reverse current sensing - fuel cell powerplants No. 1, 2, and 3
D-C main bus B	S/M jettison controller B
	D-C main bus B
Service Module Jettison Controller Battery A	
Controller A	None

Block I	Block II
Service Module Jettison Controller Battery B	
Controller B	None
Flight Bus (Powered by d-c main buses A and B)	
None	Rendezvous radar transponder
	S-band PA No. 1 transponder
	S-band power amplifier No. 2
	Up-data link
	Signal conditioning equipment
	Premodulation processor
	Data storage and S-band transmitter
	2-KMC high-gain antenna

3-45. REACTION CONTROL SYSTEM.

3-46. The reaction control system (RCS) is comprised of two subsystems: the service module and the command module reaction control systems. (See figures 3-10 and 3-11.) The primary purpose of each subsystem is to provide propulsion impulses, as required, for the accomplishment of normal and emergency attitude maneuvers of the spacecraft or the C/M. Both subsystems operate in response to automatic control signals originating from the G&N or stabilization and control system. Manual control is provided by the crew rotation hand controllers. The subsystems are similar to the extent that both utilize pressure-fed, hypergolic propellants, and maintain total redundancy of critical components and rocket engine thrust vectors.

3-47. SERVICE MODULE REACTION CONTROL.

3-48. The S/M-RCS consists of four independent, equally capable, and functionally identical packages, as shown in figure 3-10. Each package contains four reaction control engines, fuel and oxidizer tanks (one each for Block I and two each for Block II), a helium tank, and associated components such as regulators, valves, filters, and lines. These components are mounted on a panel, or package that is installed on the exterior of the S/M near the forward end. All components, with the exception of the engines, are located inside the S/M. In each package, two of the engines are for roll control, and the other two are for pitch or yaw control, depending upon the location of the package. Hypergolic propellants for the S/M-RCS consist of a 50:50 blend of UDMH and hydrazine as fuel and nitrogen tetroxide as oxidizer.

3-49. During an Apollo lunar mission, the S/M-RCS will be used to accomplish many of the following maneuvers: the service propulsion system ullage maneuver, thrust vectors for three-axis stabilization and attitude control, separation of various combinations of modules and/or boosters under normal or abort conditions, LM docking and separation, and minor orbital or midcourse velocity corrections.

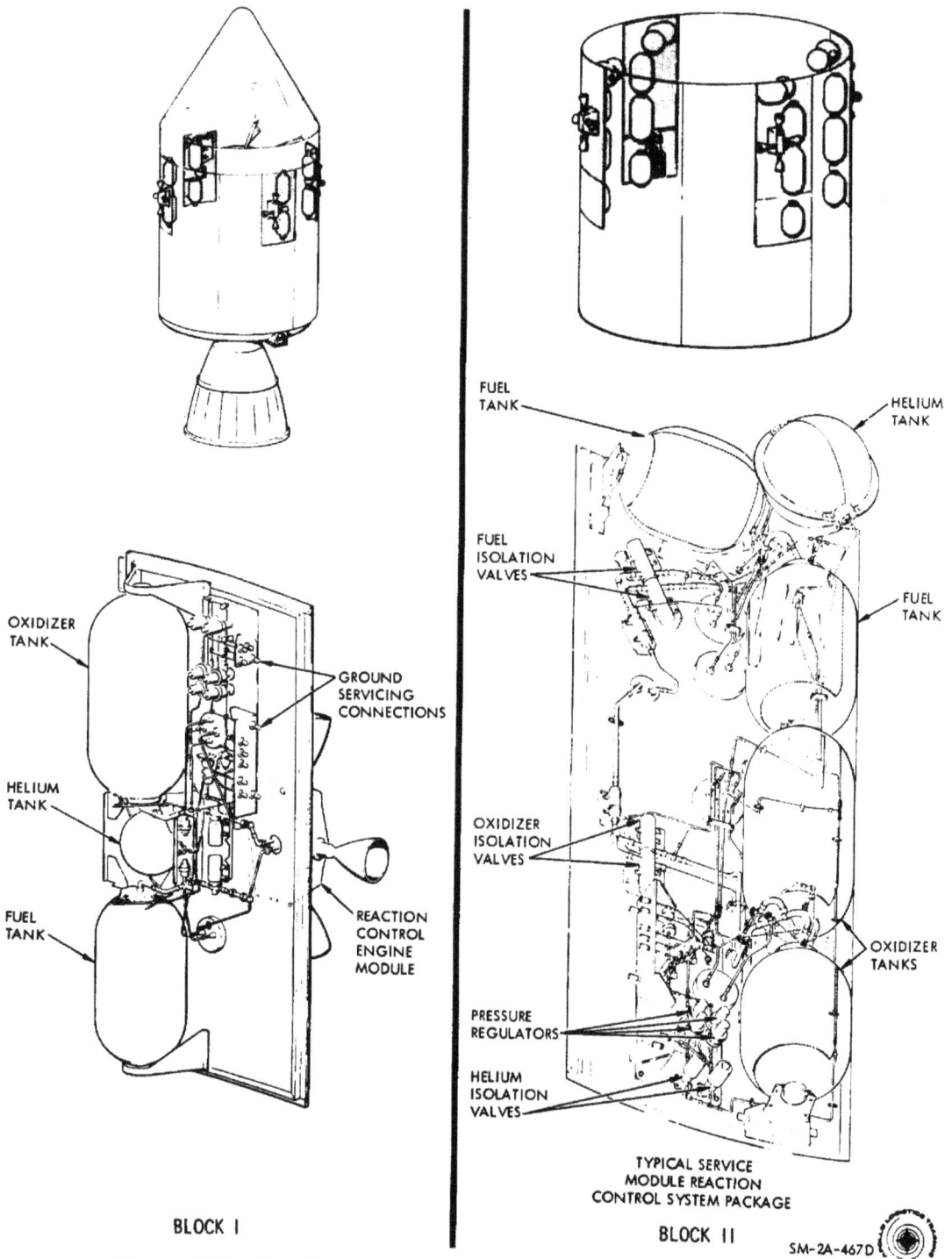

OXIDIZER
TANK

GROUND
SERVICING
CONNECTIONS

HELIUM
TANK

FUEL
TANK

REACTION
CONTROL
ENGINE
MODULE

FUEL
TANK

HELIUM
TANK

FUEL
ISOLATION
VALVES

FUEL
TANK

OXIDIZER
ISOLATION
VALVES

OXIDIZER
TANKS

PRESSURE
REGULATORS

HELIUM
ISOLATION
VALVES

TYPICAL SERVICE
MODULE REACTION
CONTROL SYSTEM PACKAGE

BLOCK I

BLOCK II   SM-2A-467D

Figure 3-10.   Service Module Reaction Control System (Sheet 1 of 2)

BLOCK II ONLY

OXIDIZER
ISOLATION
VALVE

VENT
VALVE
(LIQUID
SIDE)

OXIDIZER FILL
AND DRAIN

OXIDIZER VALVE   FUEL VALVE

PROPELLANT VALVES
& REACTION ENGINES
(4 PER PACKAGE)

VENT
VALVE
(LIQUID
SIDE)

FUEL
ISOLATION
VALVE

FUEL FILL
AND DRAIN

BLOCK II ONLY

BLADDER
(TYPICAL)

OXIDIZER TANK   FUEL TANK

VENT VALVE

VENT VALVE

BURST
DIAPHRAGM
AND RELIEF
VALVE

CHECK VALVES   CHECK VALVES

BURST
DIAPHRAGM
AND RELIEF
VALVE

REGULATOR
ASSEMBLY
NO. 1

REGULATOR
ASSEMBLY
NO. 2

HELIUM
ISOLATION
VALVE

HELIUM
ISOLATION
VALVE

HELIUM FILL
AND DRAIN
VALVE

HELIUM TANK

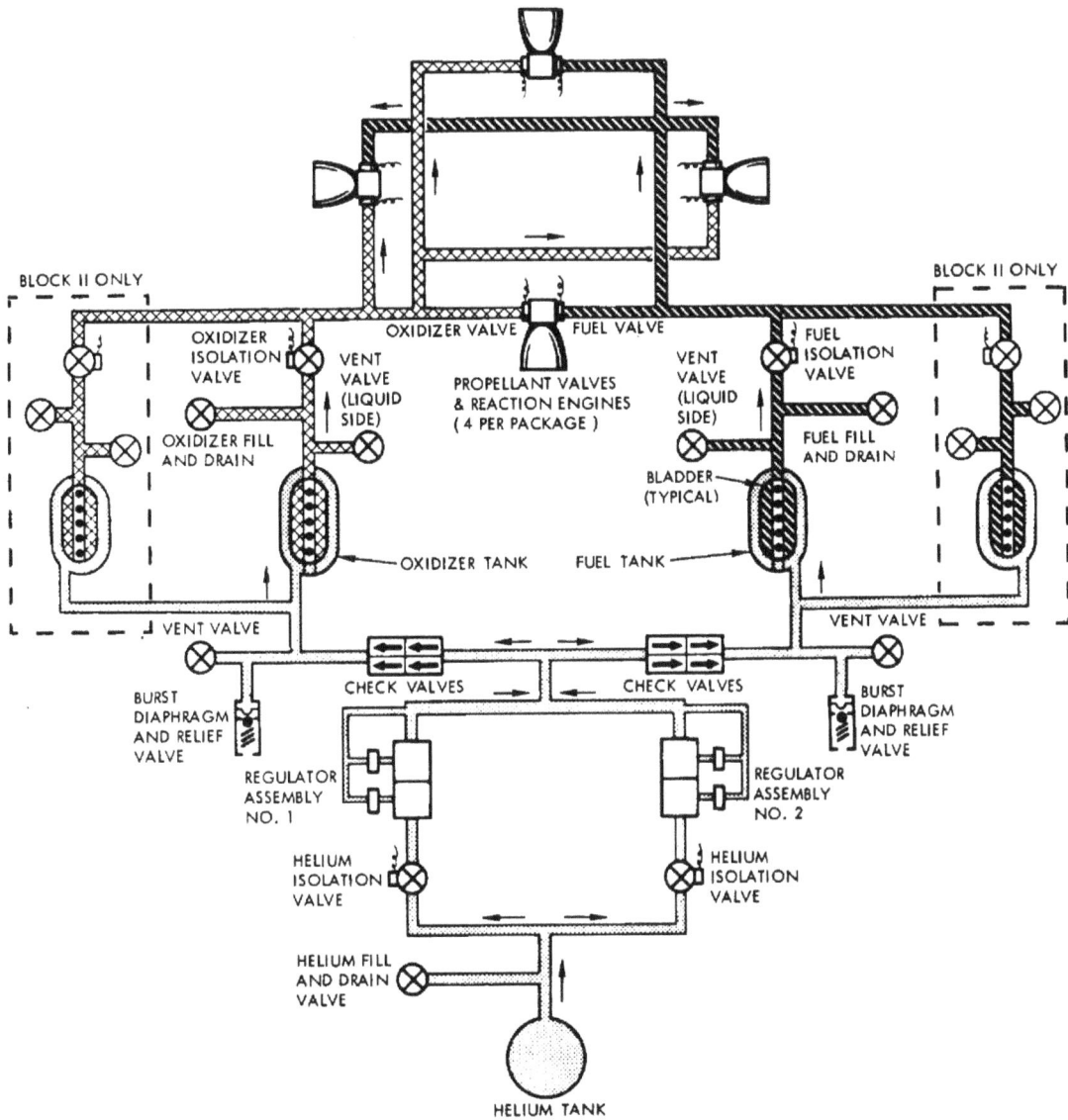

LEGEND	
▨▨▨	FUEL
⊠⊠⊠	OXIDIZER
▭▭▭	HELIUM

SM-2A-580F

Figure 3-10. Service Module Reaction Control System (Sheet 2 of 2)

3-50. <u>COMMAND MODULE REACTION CONTROL.</u>

3-51. The C/M-RCS, although similar to the S/M subsystem, is different in several respects including propellant distribution to the reaction control engines. (See figure 3-11.) The C/M contains two independent, equally capable, and functionally identical reaction control systems. Each system consists of two reaction control engines per axis (pitch, roll, and yaw), propellant storage and pressurization tanks, and associated components and lines. Hypergolic propellants for the C/M-RCS consist of nitrogen tetroxide as oxidizer, and monomethyl-hydrazine as fuel. All of the components of this subsystem are located in the aft compartment with the exception of the two negative pitch engines which are located in the forward compartment. For thrust in the pitch and yaw axes, engines from each system are mounted in pairs; whereas for roll, each pair of engines consist of a thrust left and a thrust right engine. In either case, the presence of the second system provides the required redundancy. The C/M-RCS is not activated until CSM separation takes place. The RCS is used to provide attitude maneuver capabilities during entry and, in the event of a high-altitude abort after launch escape system jettison, to provide three-axis control to aid in rate damping prior to deployment of the ELS parachutes. Due to the fact that no hypergolic propellant should be onboard the C/M at the time of earth impact, certain provisions are necessary that are not included in the S/M-RCS. Additional components, including squib-operated valves, accomplish burning of the fuel and oxidizer remaining after entry or high-altitude abort, or dump the propellant load after a pad or a low-altitude abort. On Block I the fuel is not dumped, but on Block II the total propellant load is dumped. Following either of these operations, other squib valves are activated to allow complete helium purging of the C/M-RCS fluid lines and engines.

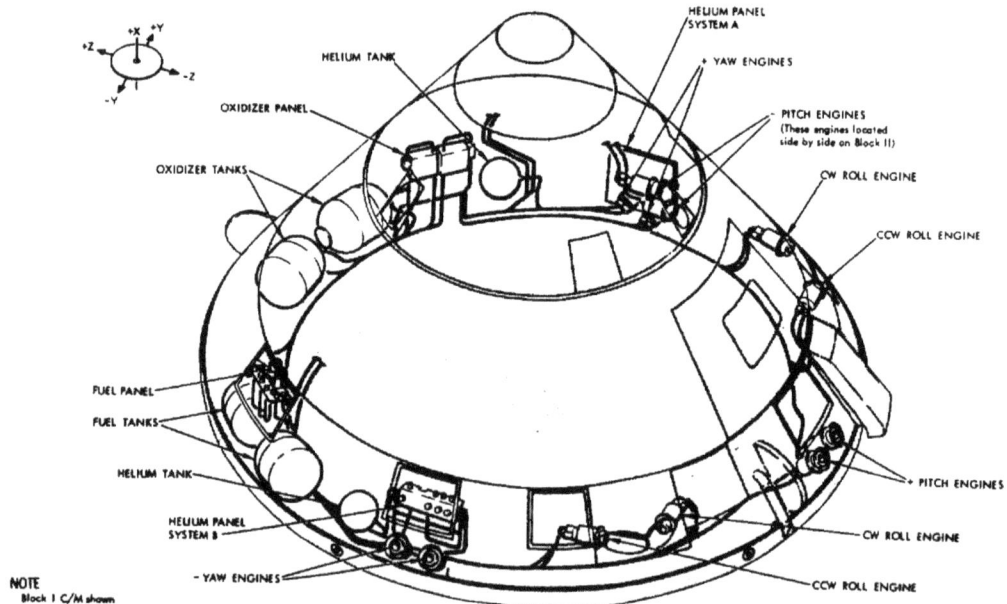

Figure 3-11. Command Module Reaction Control System (Sheet 1 of 2)

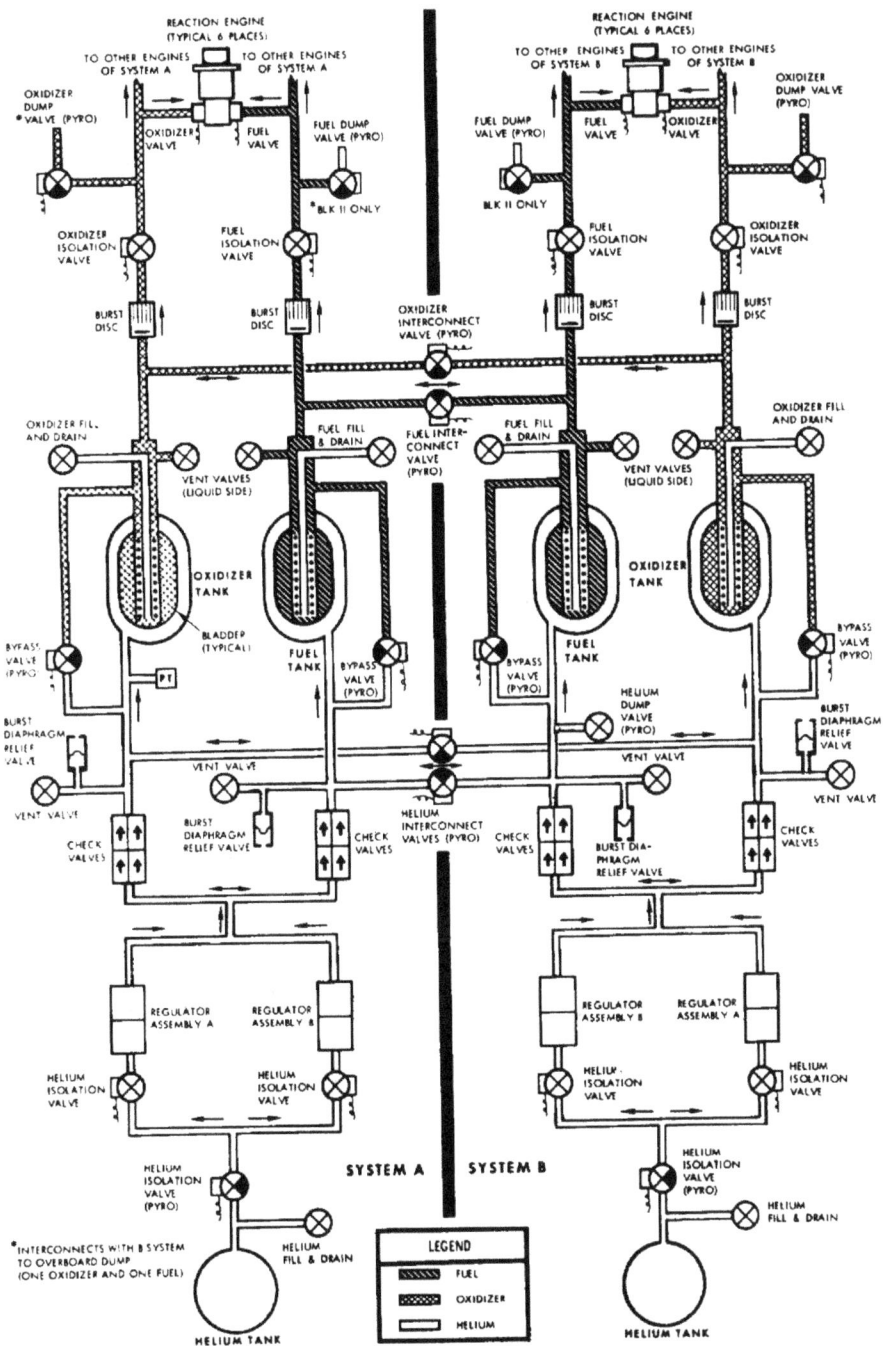

Figure 3-11. Command Module Reaction Control System (Sheet 2 of 2)

3-52. SERVICE PROPULSION SYSTEM.

3-53. The service propulsion system (SPS) provides the thrust required for large changes in spacecraft velocity after booster separation. The SPS consists of a gimbal-mounted single-rocket engine, pressurization and propellant tanks, and associated components, all of which are located in the service module. (See figure 3-12.) Conditions and time will vary the use of the SPS thrust. During a lunar landing mission, for example, the SPS could conceivably be fired for many events. The first might occur shortly after booster separation to carry out an abort during the post-atmospheric portion of the launch trajectory. It could also be used for earth orbit injection or transferring from one earth orbit to another. Following translunar injection, normal midcourse corrections or a post-injection abort could be accomplished using the SPS. Further along on the mission, the SPS provides for insertion of the S/C into lunar orbit, as well as ejection from lunar orbit (transearth injection) into a transearth trajectory. During the lunar orbit phase, transferral from one orbit to another is also possible.

3-54. Hypergolic propellants for the SPS consist of a 50:50 blend of UDMH and hydrazine as fuel, and nitrogen tetroxide as oxidizer. The storage and distribution system for these propellants consist of two fuel tanks, two oxidizer tanks, two helium tanks, associated components, lines, and electrical wiring. Pressurization of the propellant tanks is accomplished using helium. Automatic control and regulation is utilized in the SPS, as well as backup components and operational modes. A quantity gauging system is provided for monitoring the amount of propellant remaining in the tanks.

3-55. SERVICE PROPULSION SYSTEM OPERATION.

3-56. The operation of the SPS is in response to an automatic firing command generated by the G&N system, or manual initiation by the crew. The engine assembly is gimbal-mounted to allow engine thrust-vector alignment with the S/C center of gravity to preclude S/C tumbling. Thrust-vector control during firing is maintained automatically by the SCS or manually by the crew utilizing the hand controllers. The control and monitoring of the SPS from the C/M is the only point of SPS interface with the C/M. The engine has no throttle, thus it produces a single value of thrust for velocity increments.

3-57. SPS QUANTITY GAUGING AND PROPELLANT UTILIZATION SYSTEM. Service propulsion system propellant quantities are monitored and controlled by the system illustrated in figure 3-13. Propellant quantity sensing incorporates two separate systems: primary and auxiliary. The primary quantity sensors are cylindrical capacitance probes mounted axially in each tank. The auxiliary network utilizes impedance-type point sensors providing a step function impedance change when the liquid level passes their location centerline. Auxiliary system electronics provide time integration to permit a continuous measurement when the propellant level is between point sensors.

3-58. Sensor outputs are applied to a control unit containing servo loops which provide output signals representative of the propellant quantity. Two primary fuel servos and one auxiliary fuel servo provide an input signal to the fuel display servo, which, in turn, positions the digital fuel quantity display on the main display console. Two primary oxidizer servos and one auxiliary oxidizer servo provide an input to the oxidizer display servo, which, in turn, positions the digital oxidizer quantity display. The primary and auxiliary fuel and oxidizer servo outputs are also applied to the unbalance display servo, which will sense any unbalance in the remaining fuel-oxidizer quantities and display the amount of unbalance on the unbalance display dial. The propellant utilization valve assembly,

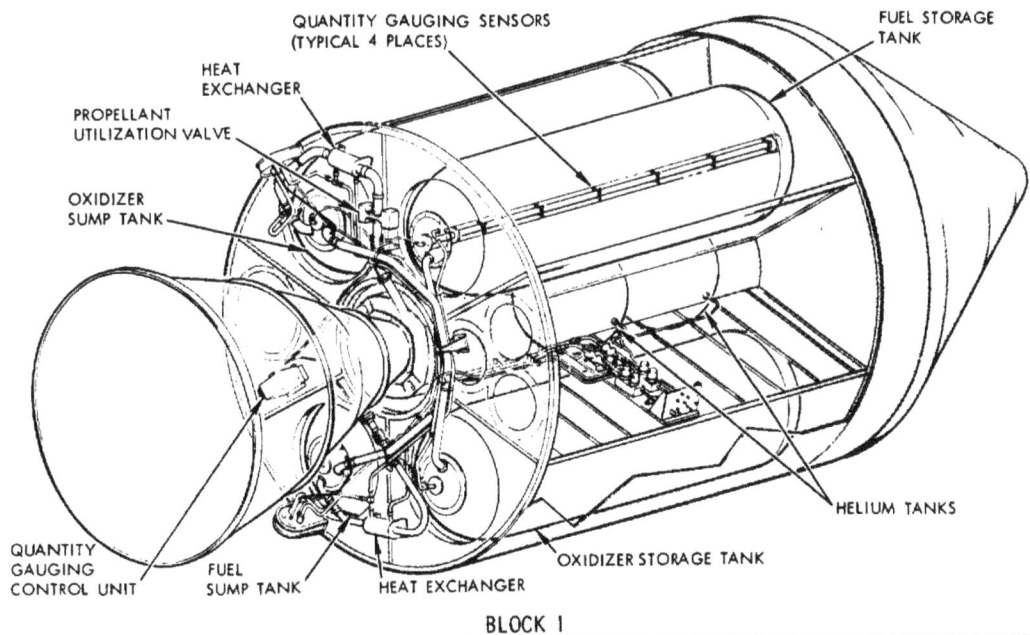

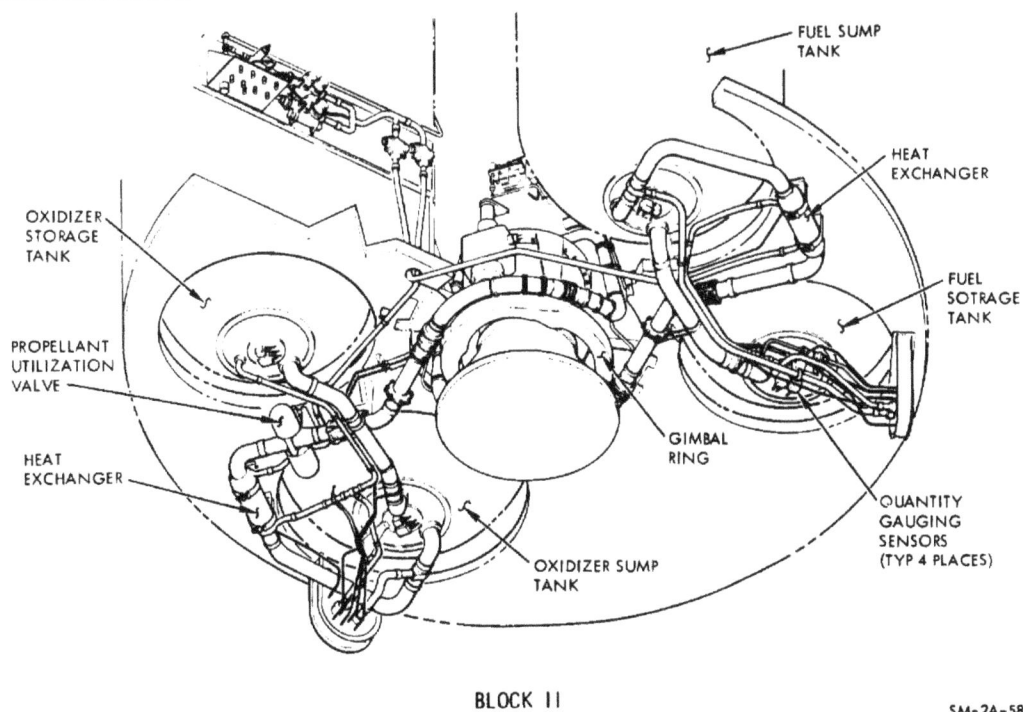

QUANTITY GAUGING SENSORS
(TYPICAL 4 PLACES)

HEAT
EXCHANGER

PROPELLANT
UTILIZATION VALVE

OXIDIZER
SUMP TANK

FUEL STORAGE
TANK

QUANTITY
GAUGING
CONTROL UNIT

FUEL
SUMP TANK

HEAT EXCHANGER

OXIDIZER STORAGE TANK

HELIUM TANKS

BLOCK I

FUEL SUMP
TANK

HEAT
EXCHANGER

FUEL
SOTRAGE
TANK

OXIDIZER
STORAGE
TANK

PROPELLANT
UTILIZATION
VALVE

HEAT
EXCHANGER

GIMBAL
RING

QUANTITY
GAUGING
SENSORS
(TYP 4 PLACES)

OXIDIZER SUMP
TANK

BLOCK II

SM-2A-582C

Figure 3-12. Service Propulsion System (Sheet 1 of 2)

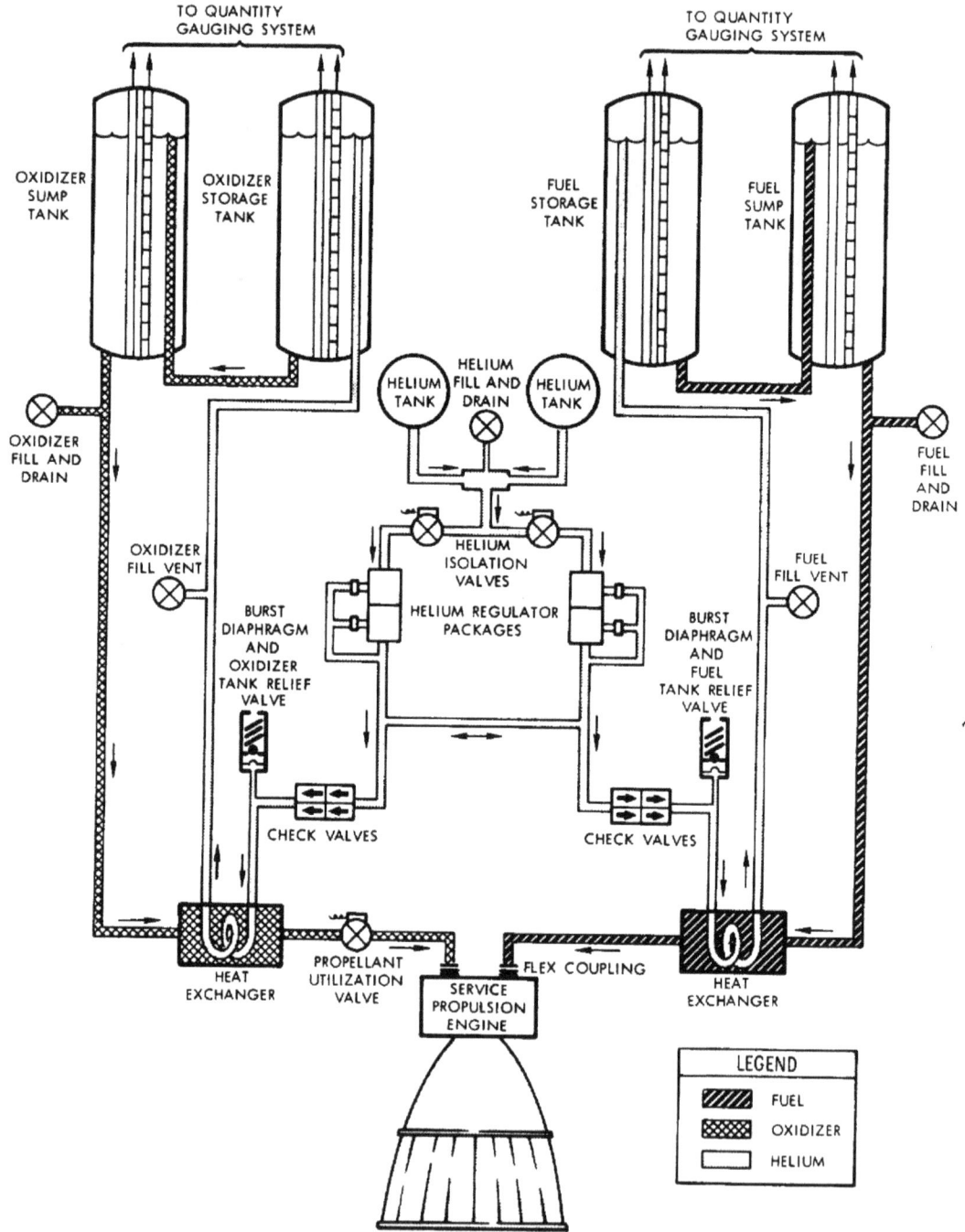

Figure 3-12. Service Propulsion System (Sheet 2 of 2)

SM-2A-469C

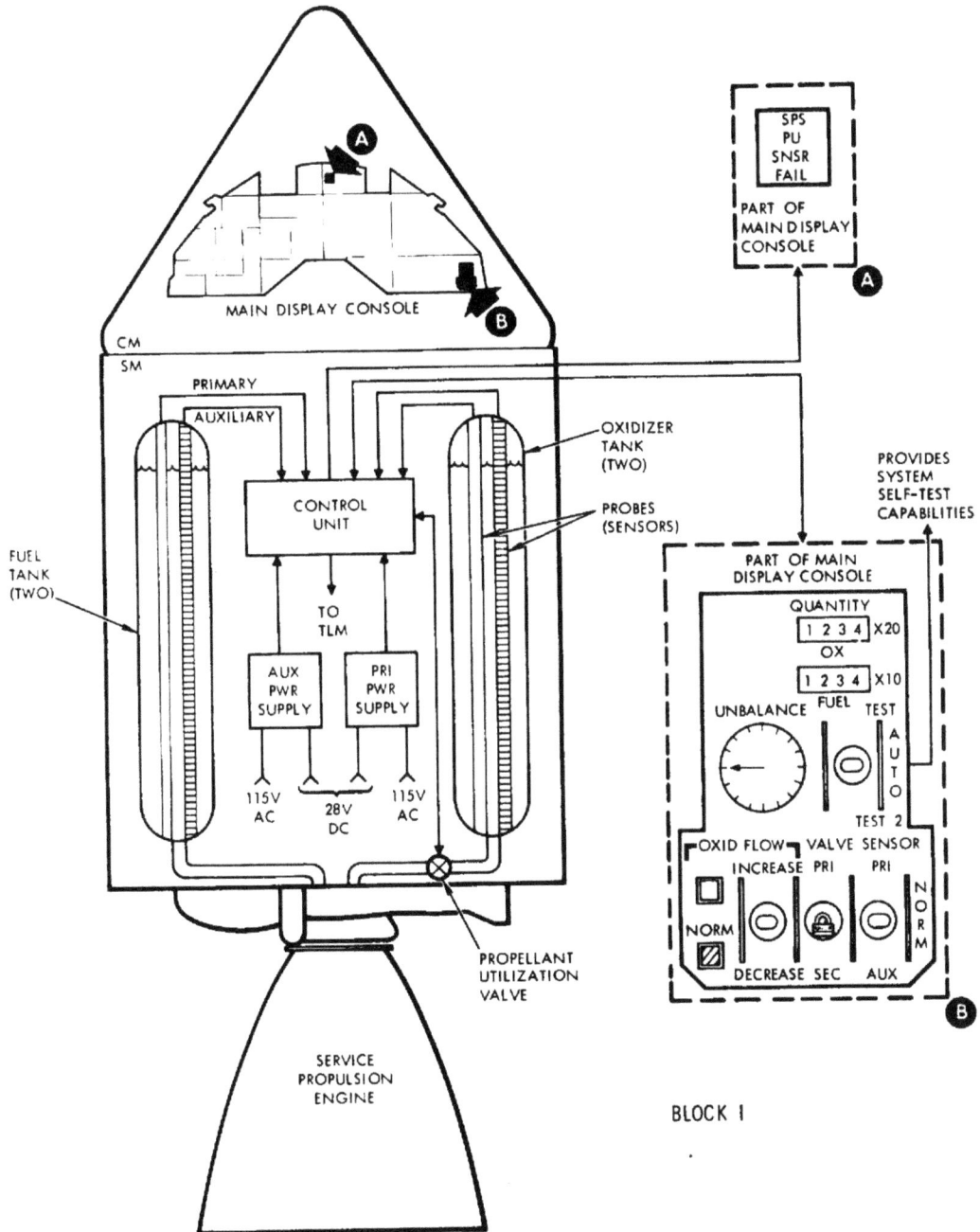

Figure 3-13. SPS Quantity Gauging and Propellant Utilization Systems — Block Diagram

SM-2A-607E

installed in the oxidizer engine feed line, incorporates two identical, motor-operated gates to provide redundant oxidizer flow rate control. The two gates (one primary and one auxiliary) are operated manually; to allow oxidizer flow rates to be increased or decreased, to compensate for an unbalance condition in the oxidizer-fuel ratio, and to insure simultaneous propellant depletion. Gate position is controlled by switches on the main display console. The valve incorporates a position potentiometer and the output is applied to a valve position display servo which positions an oxidizer flow indicator. Propellant quantity signals are also routed to telemetry. Quantity and unbalance signals are monitored by the discrepancy warning lights which will provide an alarm in the event of an excessive propellant unbalance condition. A self-test system is incorporated which provides an operational check of the sensing system voltages to the different servos, electronics, and display readings. Self-tests are initiated manually by a test switch on the main display console.

3-59. GUIDANCE AND NAVIGATION SYSTEM.

3-60. The guidance and navigation (G&N) system is a semi-automatic system, directed and operated by the flight crew, which performs two basic functions: inertial guidance and optical navigation. The system consists of inertial, optical, and computer subsystems, each of which can be operated independently, if necessary. Thus a failure in one subsystem will not disable the entire system. The G&N equipment is located in the lower equipment bay and on the main display console of the command module. (See figure 3-14.)

3-61. The three subsystems, individually or in combination, can perform the following functions:

    a. Periodically establish an inertial reference which is used for measurements and computations.

    b. Align the inertial reference by precise optical sightings.

    c. Calculate the position and velocity of the spacecraft by optical navigation and inertial guidance.

    d. Generate steering signal and thrust commands necessary to maintain the required S/C trajectory.

    e. Provide the flight crew with a display of data which indicates the status of the G&N problem.

3-62. The inertial subsystem consists of an inertial measurement unit (IMU), associated hardware, and appropriate controls and displays. Its major functions involve: (1) measuring changes in S/C attitude, (2) assisting in the generation of steering commands for the S/C stabilization and control system (SCS), and (3) measuring S/C velocity changes due to thrust. Various subsystem modes of operation can be initiated automatically by the computer subsystem or manually by the flight crew, either directly or though appropriate programing of the computer subsystem.

3-63. The optical subsystem consists of a scanning telescope, a sextant, associated hardware, and appropriate controls and displays. Its major functions involve: (1) providing the computer subsystem with data obtained by measuring angles between lines of sight to celestial objects, and (2) providing measurements for establishing the S/C inertial reference. The scanning telescope and sextant are used by the flight crew to take sightings on celestial bodies, landmarks, and the LM subsequent to separation and during rendezvous. These sightings, when used in conjunction with a catalog of celestial bodies stored in the computer subsystem, enable determination of the S/C position and orientation in space.

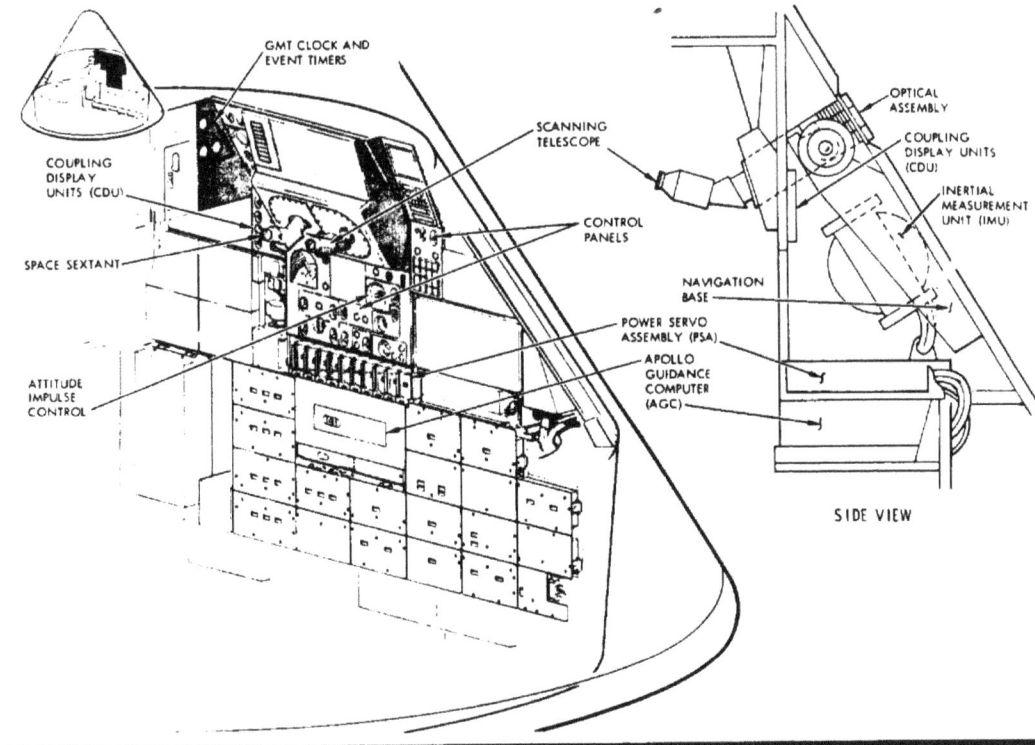

SIDE VIEW

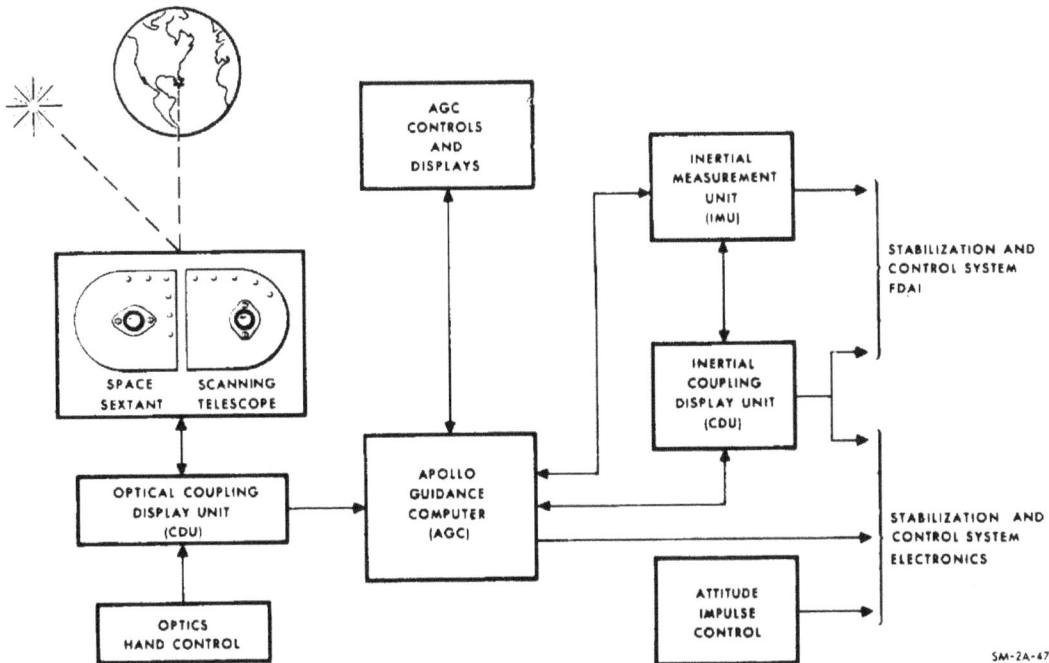

Figure 3-14. Guidance and Navigation System (Block I)

SM-2A-472G

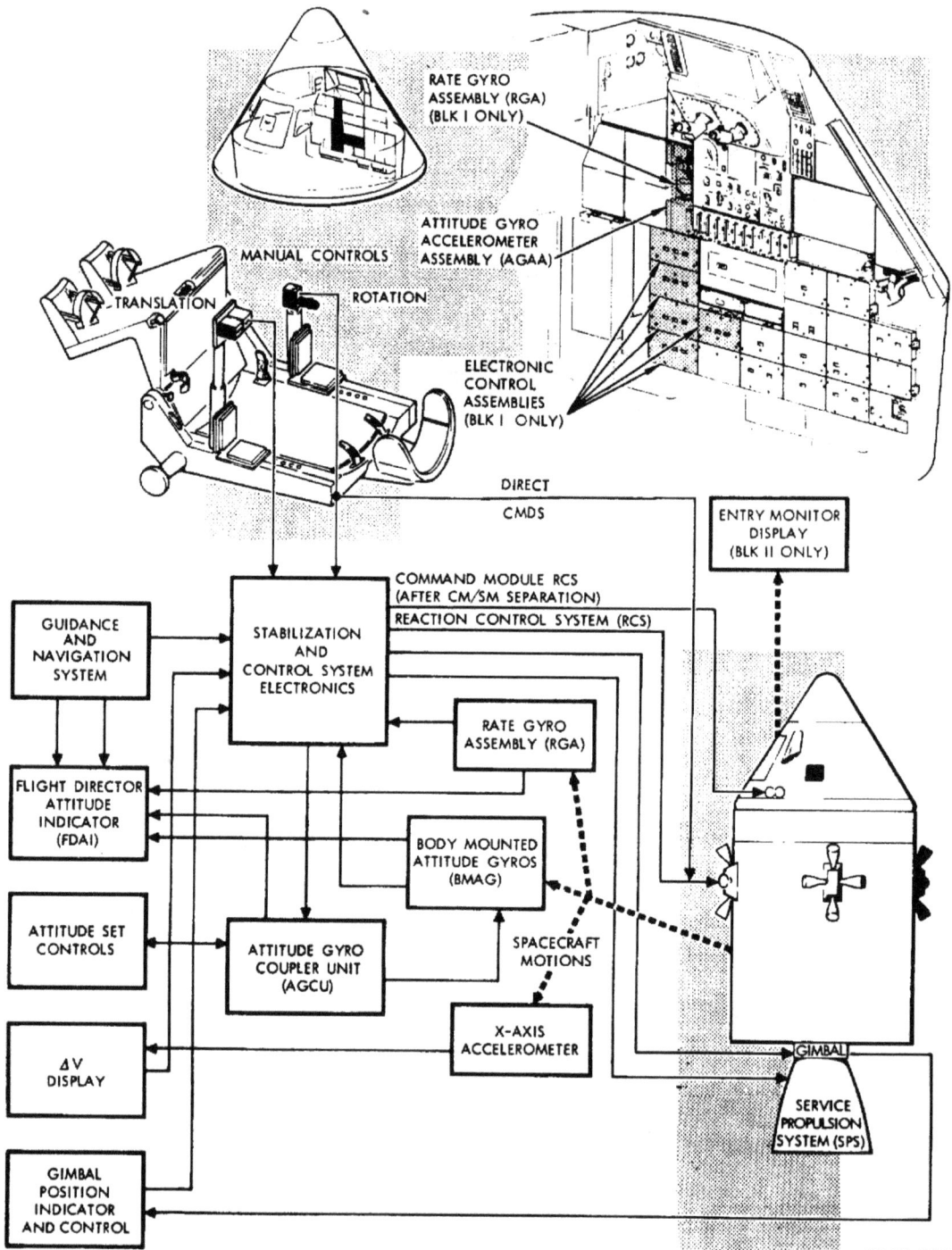

Figure 3-15. Stabilization and Control System

SM-2A-471F

Communication with ground tracking stations provides primary navigation information. The identity of celestial bodies and the schedule of measurements is based on an optimum plan determined prior to launch.

3-64. The computer subsystem consists of an Apollo guidance computer (AGC) and appropriate controls and displays. Its major functions involve: (1) calculating steering signals and discrete thrust commands to keep the S/C on a desired trajectory, (2) positioning the IMU stable platform to an inertial reference defined by optical measurements, (3) performing limited G&N system malfunction isolation, and (4) supplying pertinent S/C condition information to appropriate display panels. The AGC is a general purpose digital computer employing a core memory, parallel operation, and a built-in self-check capability. Programs are stored in the AGC and manually or automatically selected to control and solve flight equations. Using information from navigation fixes, the AGC computes a desired trajectory and calculates necessary corrective attitude and thrust commands. Velocity corrections are measured by the inertial subsystem and controlled by the computer subsystem. Velocity corrections are not made continuously but are initiated at predetermined checkpoints in the flight to conserve propellants. The G&N, SCS, SPS, and RCS systems combine to provide closed-loop control of the S/C velocity and attitude.

3-65. STABILIZATION AND CONTROL SYSTEM.

3-66. The stabilization and control system (SCS) for Block I S/C provides control and monitoring of the spacecraft attitude and rate control of the thrust vector of the service propulsion engine, and a backup inertial reference system. (See figure 3-15.) The system may be operated automatically or manually in various modes. The guidance and navigation system, service propulsion system, and the CSM reaction control system interface with the SCS. The major components of the SCS, all located in the C/M are: rate gyro assembly; attitude gyro/accelerometer assembly; pitch, yaw, and roll electronic control assemblies (ECAs); display/attitude gyro accelerometer assembly ECA; auxiliary ECA, velocity change indicator, gimbal position/attitude set indicator, flight director attitude indicator (FDAI), two rotation controls, and two translation controls. System controls and displays are located on the C/M main display console. The rate gyro assembly consists of three rate gyros mounted mutually 90 degrees apart in X-, Y-, and Z-axes. The rate gyros provide signals representative of S/C attitude change rates. The rate is displayed on the FDAI and is used by the SCS for damping and stabilization. The attitude gyro accelerometer assembly consists of three body-mounted gyros (BMAGs), and a pendulous accelerometer mounted coincident with the X-axis. The BMAGs sense pitch-, yaw-, and roll-attitude changes and provide attitude-error signals to the FDAI for display, and to the SCS for attitude control. The accelerometer provides acceleration data for automatic termination of SPS thrusting and for display on the ΔV REMAINING indicator. The ECAs are electronic modules which process and condition the input and output electrical signals of the SCS components.

3-67. STABILIZATION AND CONTROL SYSTEM OPERATION.

3-68. The SCS may be used in any one of eight modes, which are selectable by the crew. The SCS attitude control mode automatically maintains S/C attitude within the minimum or maximum deadband limits, which are approximately ±0.5 degrees or ±5.0 degrees respectively. When the S/C exceeds the selected deadband limits, attitude-error signals generated by the BMAGs are applied to circuitry within the ECAs which initiates firing of the proper RCS engines to return the S/C within the selected deadband limits. During G&N attitude control mode, the attitude error consists of the difference between the inertial measurement unit (IMU) and coupling display unit (CDU) output signals. The SCS local vertical mode operation is identical to the SCS attitude control mode with one exception. During local vertical mode operation, an orbit rate signal is generated in the attitude gyro coupling unit to maintain local vertical reference with respect to the earth. The G&N ΔV mode is the primary mode when S/C velocity changes are required. The SPS thrust on-off signal is applied to the SCS electronics by the Apollo guidance computer (AGC). In the G&N mode, automatic thrust vector control (TVC) is accomplished with the G&N attitude error signals, the SCS rate gyro signals, and the TVC electronics. In the SCS ΔV mode, SPS thrust-on signal is initiated manually by the crew. Thrust vector control is accomplished with the attitude error signals generated by the BMAGs, the SCS rate gyro signals, and the TVC electronics. Termination of SPS thrusting is automatic and occurs when the X-axis accelerometer senses that the desired velocity change and the ΔV REMAINING indicator reads zero. Thrust on or off may be accomplished manually if automatic on or off functions fail to occur. Manual control of the SPS thrust vector with rate damping is available to the crew. The translation control is rotated CW for manual control (MTVC) and the rotation control is used to command the SPS engine in the pitch and yaw axis. During entry the G&N entry mode is the primary mode. After CSM separation, the G&N system provides automatic control of the S/C lift vector to effect a safe entry trajectory. In the SCS entry mode, the crew is required to manually control the S/C lift vector with the rotation controller. The SCS is considered a backup system during ΔV maneuvers and entry. The monitor mode permits the crew to monitor S/C attitude, attitude error, and attitude change rate on the FDAI during ascent, and provides rate stabilization after S-IVB separation.

3-69. SPACECRAFT CONTROL PROGRAMERS.

3-70. CONTROL PROGRAMER.

3-71. The control programer (M1), installed in S/C 009, automatically controlled the stabilization control system (SCS) and mission sequencer, and conditioned ground command signals. The control programer consisted of a timer assembly, radio command control assembly, and automatic command control assembly. Equipment required with the control programer consisted of a backup attitude reference system (ARS), and radio command equipment. Redundancy was provided to preclude the possibility of losing a C/M due to a single failure. A functional block diagram of the control programer is shown in figure 3-16; the mission description for S/C 009 is presented in section VII.

3-72. MISSION CONTROL PROGRAMER.

3-73. The mission control programer (M3), installed in S/C 011, 017, and 020 programs spacecraft functions which are commanded by the G&N system, S-IVB instrument unit, and the up-data link. The guidance and control (G&C) functions are controlled by the G&N system. The S/C will have operational controls and displays. Characteristics incorporated in the mission control programer include relay switching of automated functions (relay logic) and fixed time-delay relays as required, minimum interference with S/C control and display operation, real time of control of functions computed for C/M recovery, and redundant critical functions. A functional block diagram of the mission control programer is shown in figure 3-16; the mission description for S/C 011 is presented in section VII.

SPACECRAFT 009 CONTROL PROGRAMER

SPACECRAFT 011 MISSION CONTROL PROGRAMER

SM-2A-626B

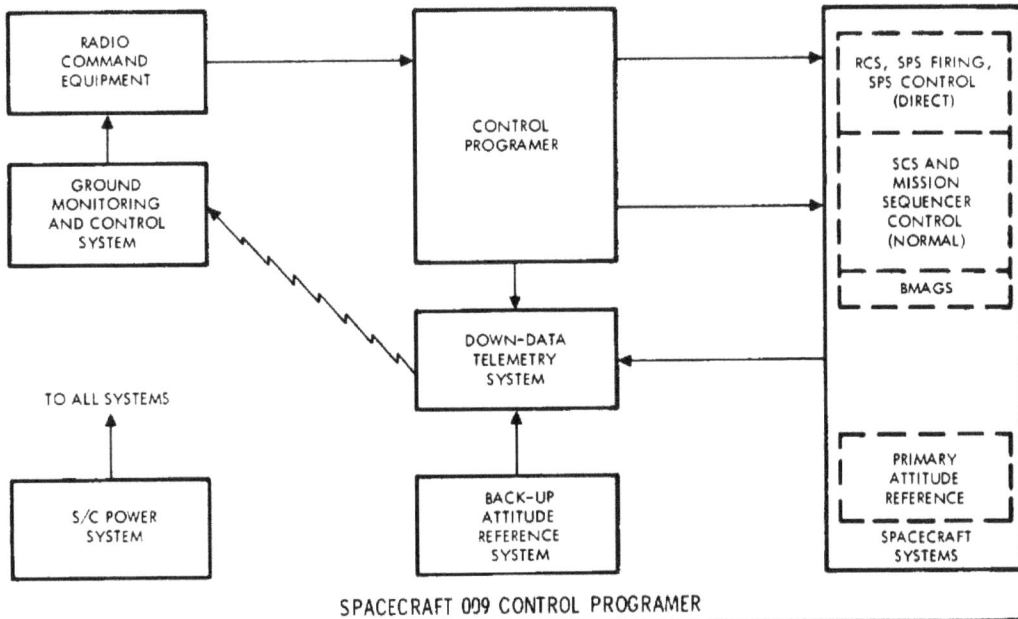

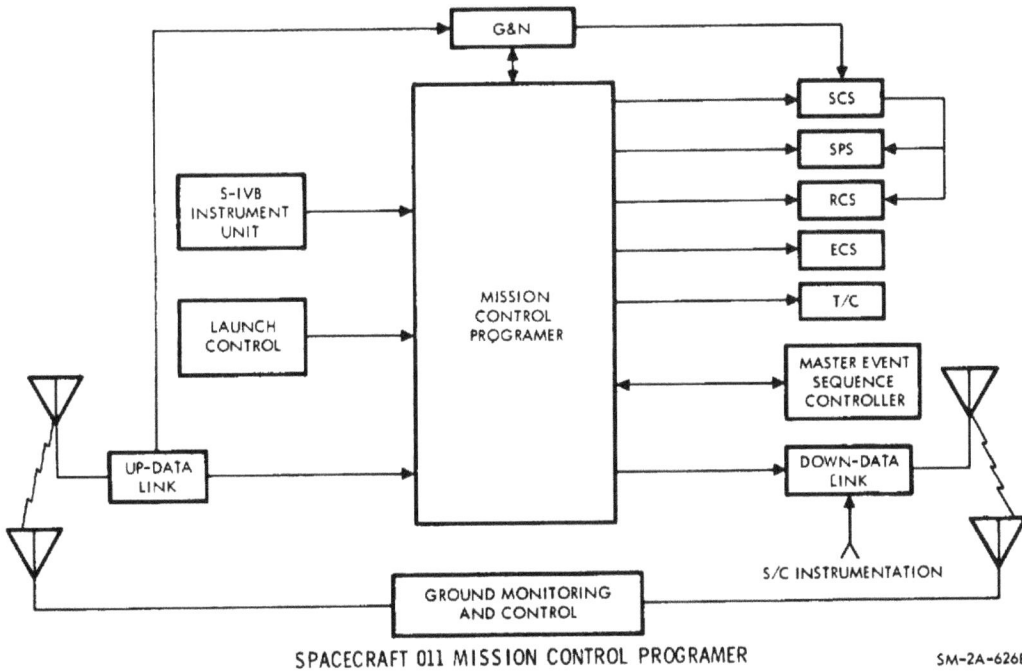

Figure 3-16. Control Programer Functional Block Diagram
for S/C 009 and S/C 011

3-74. SPACECRAFTS 009 and 011 PROGRAMER COMPARISON.

3-75. The following list provides a functional comparison of the control programers used in S/C 009 and S/C 011:

Spacecrafts 009 and 011 Programer Comparison

Functions	Control Programer (S/C 009)	Mission Control Programer (S/C 011)
Mission director ·	Primary	None
S/C system sequencing	Primary	Initiated by G&N and S-IVB instrument unit
Attitude reference	SCS (Primary) and attitude reference system (backup)	G&N (Primary) and SCS (backup)
Corrective action	EPS only	Limited
Ground control	G&C and staging	Attitude control and staging
Abort capability (maneuvers)	Self-contained and ground control	Ground control
Displays and controls	Partial	Complete and operational

3-76. CREW SYSTEM.

3-77. The purpose of the crew system is to provide for needs peculiar to the presence of the crew aboard the spacecraft. Crew system equipment includes certain degrees of physical protection against acceleration, impact, and sustained weightlessness. Equipment and provisions for the routine functions of eating, drinking, sleeping, body cleansing, and the elimination of waste are part of the crew system, as is the survival equipment which is provided for abnormal or emergency conditions.

3-78. CREW COUCHES.

3-79. The command module is equipped with three couches, each with adjustable headrests, restraint harness assemblies, seats, and foot-strap restraints. (See figure 3-17.) The basic support for all couches is a fixed frame suspended from shock attenuators. Angular adjustments for hips permit individual comfort during all flight modes. The attenuators lessen the impact forces imposed on the crew during C/M touchdown on water or land.

3-80. PERSONAL EQUIPMENT.

3-81. Each astronaut will have personal equipment available to him during the course of a 14-day mission. The equipment includes a communications (soft hat) assembly, a constant-wear garment, a pressure garment assembly (pressure suit), an umbilical assembly, a bioinstrument accessories kit, radiation dosimeters, an emergency medical kit, and a physiological clinical monitoring instrument set. In addition, a thermal insulation overgarment and a portable life support system (PLSS) will be included with the Block II spacecraft.

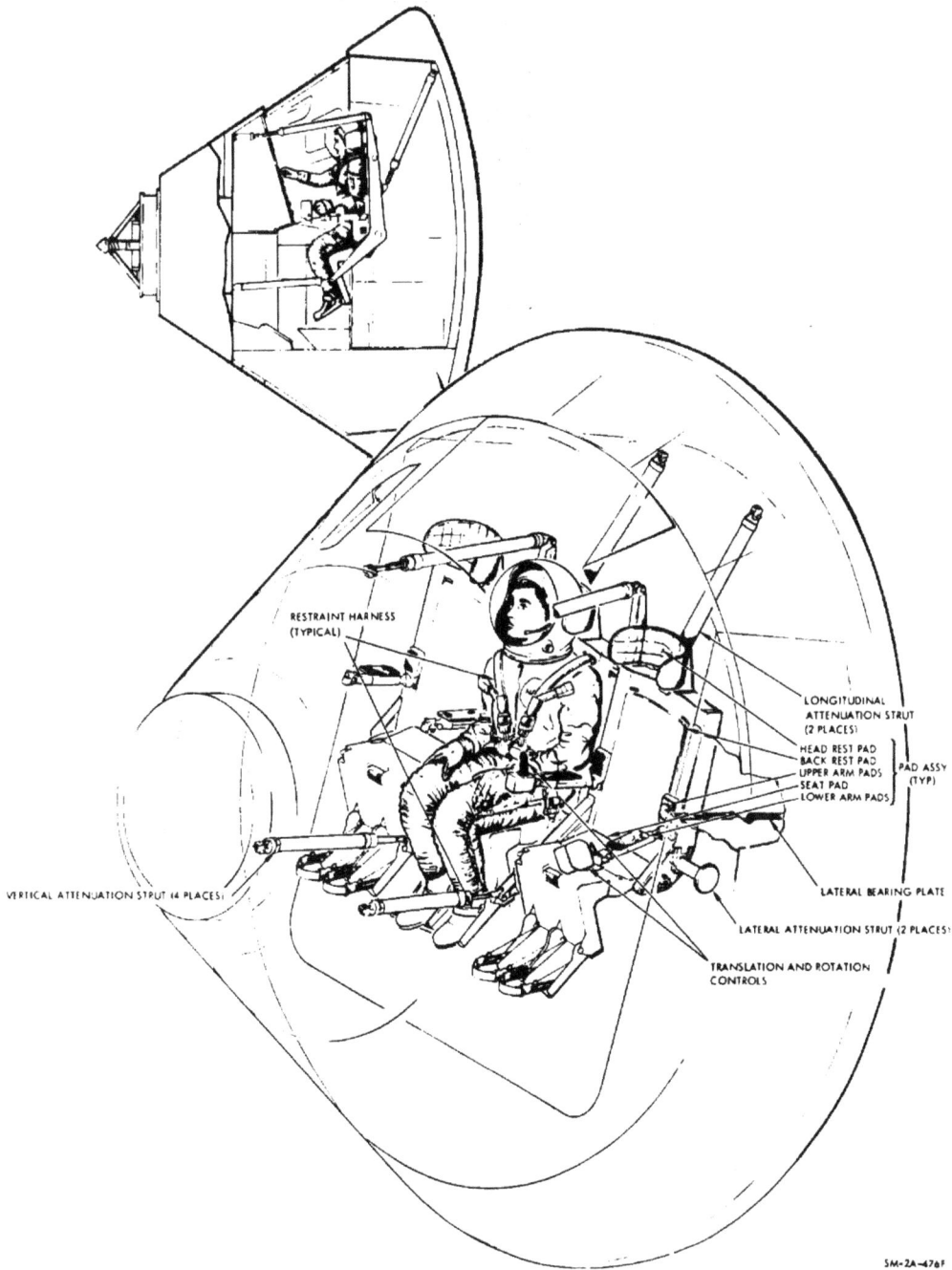

RESTRAINT HARNESS
(TYPICAL)

LONGITUDINAL
ATTENUATION STRUT
(2 PLACES)

HEAD REST PAD
BACK REST PAD
UPPER ARM PADS
SEAT PAD
LOWER ARM PADS

PAD ASSY
(TYP)

VERTICAL ATTENUATION STRUT (4 PLACES)

LATERAL BEARING PLATE

LATERAL ATTENUATION STRUT (2 PLACES)

TRANSLATION AND ROTATION
CONTROLS

SM-2A-476F

Figure 3-17. Crew Couches and Restraint Equipment

pad assembly is installed on each crew couch for crewman comfort. Restraint harness assemblies and foot-strap restraint assemblies are installed on each crew couch primarily to restrain the crewmen during critical phases of the mission. Restraint sandals, worn over the feet of constant-wear garments, have soles made of Velcro pile material, which adheres to Velcro hook material installed on the C/M floor, and parts of structural surface.

3-94. WASTE MANAGEMENT SYSTEM.

3-95. The waste management segment of the crew system consists of the means for urine disposal, collecting and storing fecal matter and personal hygienic wastes. Fecal matter is collected in plastic (polyethylene) bags, disinfected, and stored in a compartment. Personal hygienic wastes are also collected and stored in this manner. The urine is expelled overboard by differential pressure method. This is accomplished by properly positioning two manually controlled waste management valves. (See figure 3-18.) A vent line interfaces the WMS to the WMS overboard dump line, and provides for the removal of odors originating as a result of waste management functions.

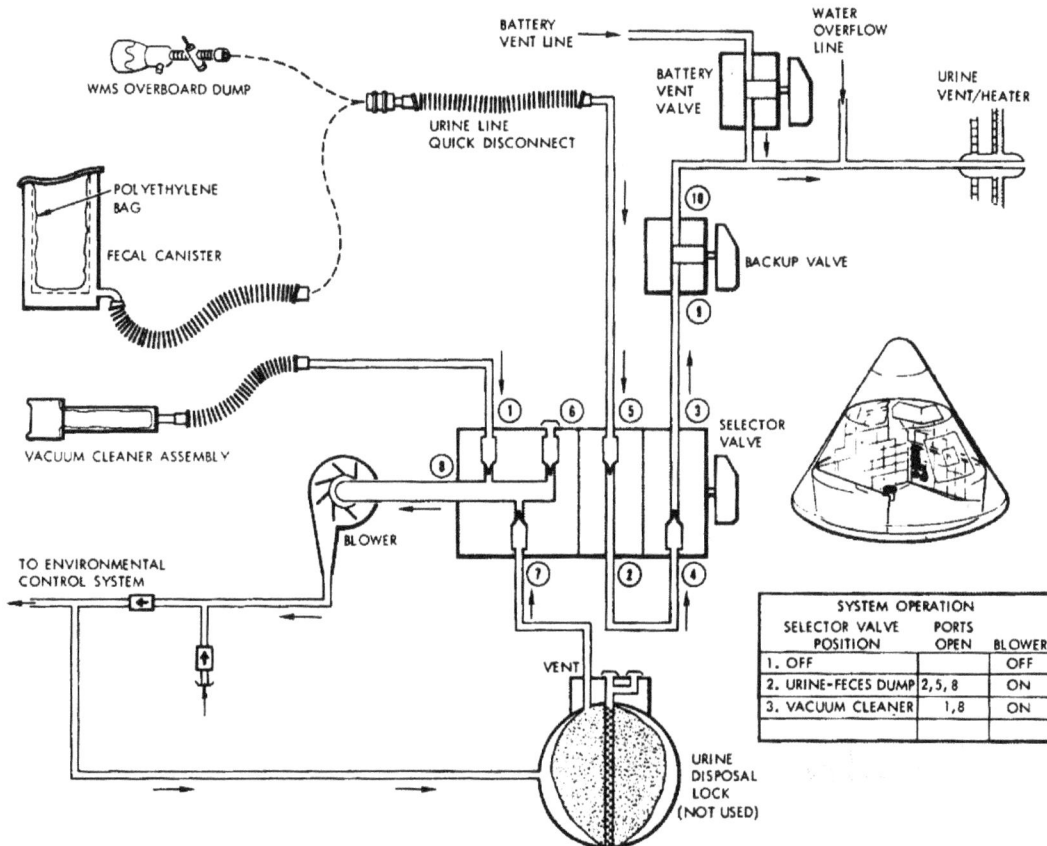

SYSTEM OPERATION		
SELECTOR VALVE POSITION	PORTS OPEN	BLOWER
1. OFF		OFF
2. URINE-FECES DUMP	2, 5, 8	ON
3. VACUUM CLEANER	1, 8	ON

SM-2A-477E

Figure 3-18. Waste Management System Functional Diagram

3-82. The communications (soft hat) assembly provides communications between crewmembers and MSFN. The assembly consists of a adjustable strap helmet with two microphones and two earphones attached, and will be used in a shirtsleeve environment. During Block II missions, the communications soft hat will be worn in the PGA helmet also.

3-83. The constant-wear garment (CWG) provides the astronauts with a basic garment to be worn at all times during a mission. The CWG is a one-piece short-sleeved, close-fitting garment which covers the crewman's entire body with the exception of the arms and head.

3-84. The thermal insulation overgarment coverall is an overgarment which completely covers the pressure garment and PLSS. It provides thermal and micrometeriod protection for the astronauts during extravehicular operations.

3-85. The pressure garment assembly (PGA) consists of a torso-and-limbs covering, integral boots, gloves, and helmet, and covers the entire body of an astronaut. The garment is air-tight to support life, in conjunction with the ECS or PLSS, during adverse mission conditions and lunar exploration.

3-86. The umbilical assembly provides the interface between the S/C systems and the PGA. The electrical link between the astronauts and the S/C communications equipment is provided by the electrical umbilical. The oxygen umbilical provides a means of transferring oxygen from the ECS to the PGA. Another hose assembly, provided on Block II manned S/C, is used to transfer oxygen from the ECS to the PLSS.

3-87. The biomedical harness consists of biomedical sensors, wire harness, and biomedical preamplifiers. Purpose of the sensors is to acquire the electrical signal required to determine the respiration rate and electrocardiograms of an astronaut. The preamplifier is used to condition and relay the signals received by the sensors to the telemetry system for transmittal to earth. A biomedical accessories kit is aboard to replace faulty biomedical harness components.

3-88. The radiation dosimeters measure and record the amount of radiation to which the astronaut is exposed. One is located at the right temple of the constant-wear communications assembly; others are located in pockets provided on constant-wear garments.

3-89. The emergency medical kit provides the equipment and medications required for emergency treatment of illness or injuries sustained by crewmen during a mission.

3-90. The physiological clinical monitoring instrument set consists of an aneroid sphygmomanometer, a stethoscope, and a thermometer, and is used to measure blood pressure, respiration rate or heart beat, and body temperature.

3-91. The portable life support system is a small self-contained environmental control system and communications unit, and provides life support during extravehicular activities and lunar surface exploration. The PLSS is worn as a backpack and is capable of maintaining the PGA in a pressurized condition for 4 hours without recharging.

3-92. CREW COUCH AND RESTRAINT EQUIPMENT.

3-93. Crew couch and restraint equipment consists of crew couch pad assemblies, restraint harness assemblies, foot-strap restraint assemblies, and restraint sandals. A

3-96. CREW SURVIVAL EQUIPMENT.

3-97. Two survival kits are stowed in the C/M and are available to the crew during the postlanding phase (water or land) of a mission. The major items provided include a container with 5 pounds of water, a desalter kit, three one-man liferafts (one three-man liferaft is provided in Block II C/Ms), a radio-beacon, portable light, sunglasses, machete (and sheath), and a medical kit. The liferaft includes additional equipment such as a sea anchor, dye marker, sunbonnet, etc.

3-98. FOOD, WATER, AND ASSOCIATED EQUIPMENT.

3-99. Adequate food, water, and personal hygiene aids will be provided for the total length of the mission. Small polyethylene bags containing "freeze dried" food will be stored in the C/M. By adding water and kneading, the food mixtures can be squeezed into the crewmember's mouth. Either hot or cold water is available at the potable water supply panel for food reconstitution. Chilled drinking water will be supplied to the astronauts by a single flexible hose assembly from the water delivery unit. This water source, a by-product of the fuel cell powerplants, will furnish the crew up to 36 pounds (17 quarts) of water per day. A folding shelf is provided as a convenient surface for tools, food packages, and equipment. Personal hygiene aids consist of items for oral hygiene and body cleansing such as chewing gum, interdental stimulators, and cleansing pads.

3-100. CREW ACCESSORIES.

3-101. Several accessory items are considered to be part of the crew system. These consist of the following: in-flight tool set, extra vehicular crew transfer mechanism, mirror assemblies for increased interior and exterior vision, main display console handhold straps, and an optical alignment sight. The handhold straps are installed as an aid to crewman mobility within the C/M, and the optical alignment sight is utilized to properly orient and align the CSM with the LM while accomplishing docking maneuvers.

3-102. COMMAND MODULE INTERIOR LIGHTING.

3-103. Block I C/M interior lighting equipment (figure 3-19) provides light for the main control panels in the command module and consists of eight floodlight fixture assemblies and three control panels. Each fixture assembly contains two fluorescent lamps (one primary and one secondary) and a converter. The interior lighting is powered by d-c main buses A and B, assuring a power source for lights in all areas in the event that either bus fails. The converter in each floodlight fixture converts 28 volts dc to a-c power to operate the fluorescent lamps. The floodlights are used to light three areas: the main display console (left and right areas) and the LEB area.

3-104. Each control panel has a primary and secondary control for the floodlights in its respective area. The primary control is a rheostat that controls brightness of the primary floodlights. The secondary control is an on-off switch for the secondary floodlights and is set to on when additional brightness is desired.

3-105. Interior lighting for Block II S/C (figure 3-19) is essentially the same as Block I S/C, with the addition of electroluminescent lighting to the control and display panels.

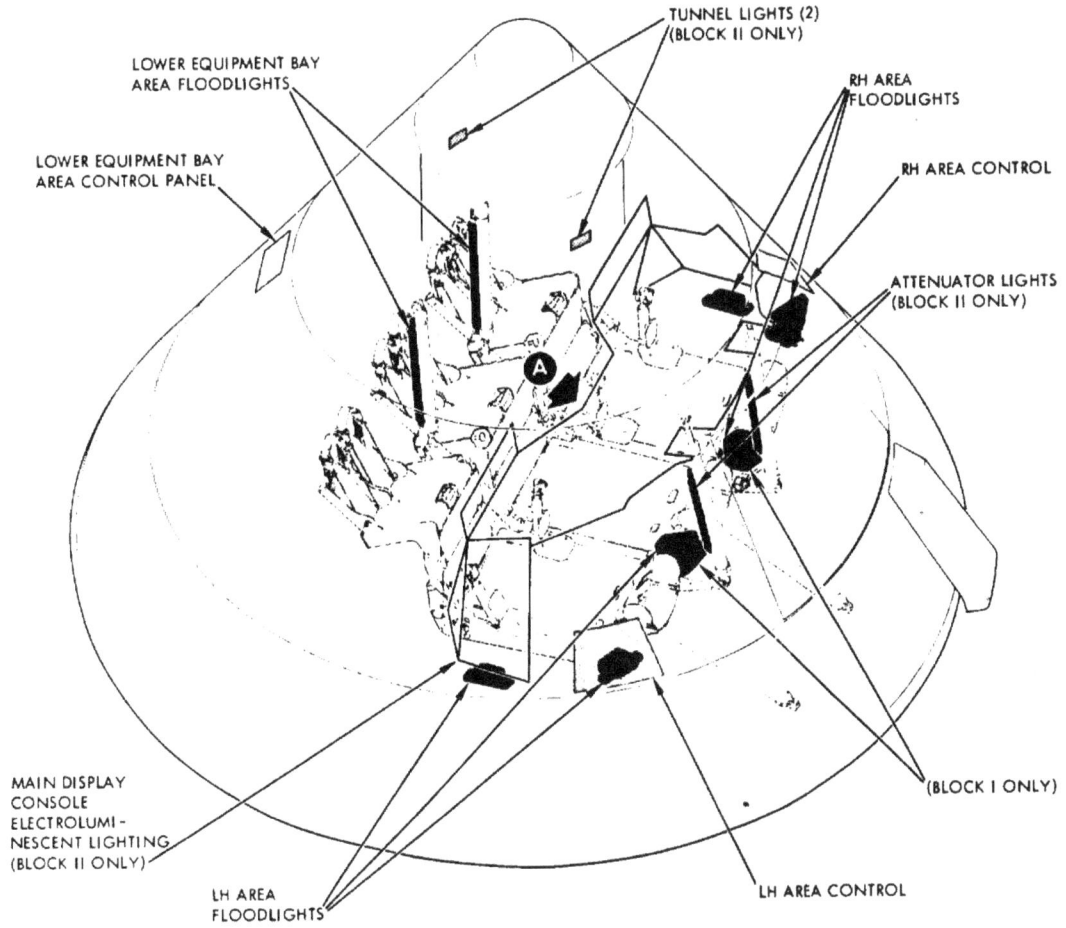

TUNNEL LIGHTS (2)
(BLOCK II ONLY)

LOWER EQUIPMENT BAY
AREA FLOODLIGHTS

RH AREA
FLOODLIGHTS

LOWER EQUIPMENT BAY
AREA CONTROL PANEL

RH AREA CONTROL

ATTENUATOR LIGHTS
(BLOCK II ONLY)

A

(BLOCK I ONLY)

MAIN DISPLAY
CONSOLE
ELECTROLUMI-
NESCENT LIGHTING
(BLOCK II ONLY)

LH AREA
FLOODLIGHTS

LH AREA CONTROL

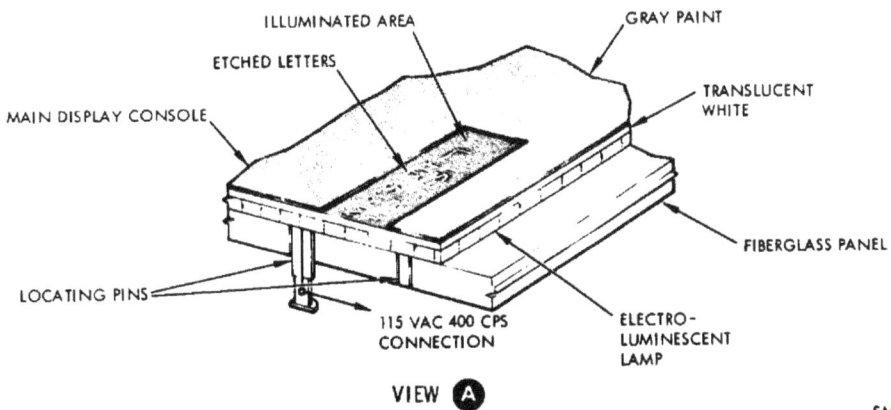

ILLUMINATED AREA

GRAY PAINT

ETCHED LETTERS

TRANSLUCENT
WHITE

MAIN DISPLAY CONSOLE

FIBERGLASS PANEL

LOCATING PINS

115 VAC 400 CPS
CONNECTION

ELECTRO-
LUMINESCENT
LAMP

VIEW A

SM-2A-851A

Figure 3-19. Command Module Interior Lighting Configuration

3-106. TELECOMMUNICATION SYSTEM.

3-107. The function of the telecommunication system (figure 3-20) is to provide for the communication of voice, television, telemetry, and tracking and ranging data between the S/C and the MSFN, the LM and EVA PLSS. It also provides for S/C intercommunications and, includes the central timing equipment for synchronization of other equipment and correlation of telemetry data. The T/C system contains the following equipment, listed in four groups:

a. Data equipment group

   - Signal conditioning equipment (SCE)
   - Pulse code modulation/telemetry (PCM/TLM) equipment
   - Television (TV) equipment
   - Up-data link (UDL) equipment
   - Premodulation processor (PMP) equipment
   - Data storage equipment (DSE)
   - Flight qualification recorder (FQR)
   - Central timing equipment (CTE)

b. Intercommunications equipment group

   - Audio center equipment
   - Headsets and connecting electrical umbilicals

c. RF electronics equipment group

   - VHF/AM transmitter - receiver equipment (2) (Block II)
   - VHF/FM transmitter equipment (Block I)
   - HF transceiver equipment
   - VHF recovery beacon equipment
   - Unified S-band equipment (USBE)
   - S-band power amplifier (S-band PA) equipment
   - C-band transponder equipment
   - Rendezvous radar transponder equipment (Block II only)

d. Antenna equipment group

   - VHF/2-KMC omni-antenna equipment
   - 2-KMC high-gain antenna equipment (Block II only)
   - VHF recovery antenna equipment
   - HF recovery antenna equipment
   - C-band beacon antenna equipment
   - Rendezvous radar antenna equipment (Block II only)

Controls and switches for operation of the T/C system are located near the pilots station in the crew compartment. Also, there are three separate groups of controls (one for each crewmember) for individual control of audio inputs and outputs of the crewmembers headsets.

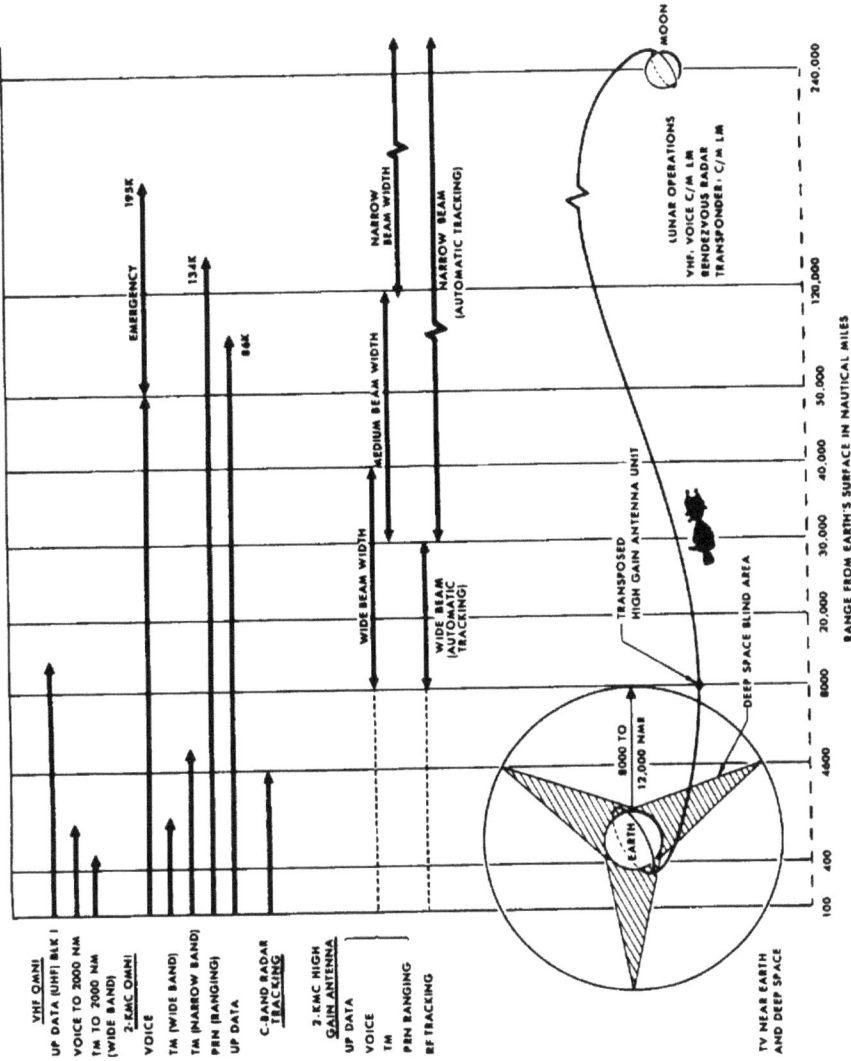

COMMUNICATION RANGES

VHF OMNI
UP DATA (UHF) BLK I
VOICE TO 2000 NM
TM TO 2000 NM
(WIDE BAND)

2-KMC OMNI
VOICE
TM (WIDE BAND)
TM (NARROW BAND)
PRN (RANGING)
UP DATA

C-BAND RADAR
TRACKING

2-KMC HIGH
GAIN ANTENNA
UP DATA
VOICE
TM
PRN RANGING
RF TRACKING

TV NEAR EARTH
AND DEEP SPACE

EMERGENCY   195K
134K
86K

WIDE BEAM WIDTH

MEDIUM BEAM WIDTH   NARROW
BEAM WIDTH

WIDE BEAM
(AUTOMATIC
TRACKING)

NARROW BEAM
(AUTOMATIC TRACKING)

TRANSPOSED
HIGH GAIN ANTENNA UNIT

DEEP SPACE BLIND AREA

8000 TO
12,000 NME

EARTH

100   400   4000   8000   20,000   30,000   40,000   50,000   120,000   240,000

RANGE FROM EARTH'S SURFACE IN NAUTICAL MILES

LUNAR OPERATIONS
VHF, VOICE C/M LM
RENDEZVOUS RADAR
TRANSPONDER C/M LM

MOON

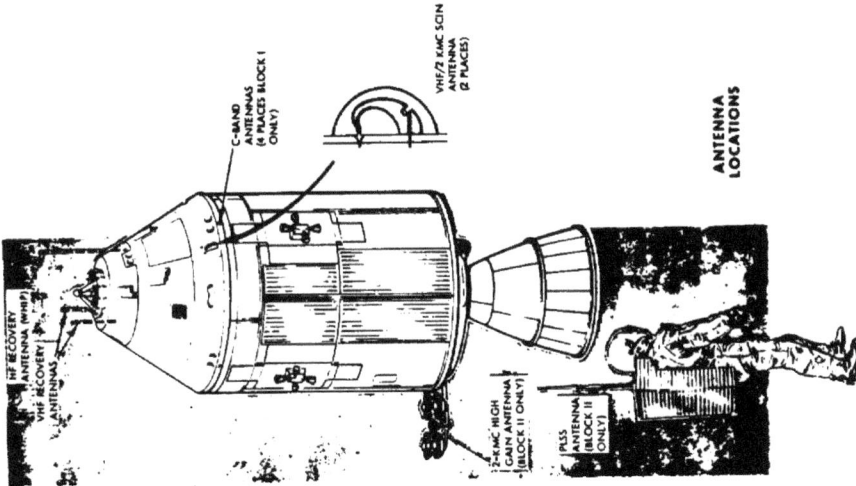

HF RECOVERY
ANTENNA (WHIP)

VHF RECOVERY
ANTENNAS

C-BAND
ANTENNAS
(4 PLACES BLOCK I
ONLY)

VHF/2 KMC SCIN
ANTENNA
(2 PLACES)

2-KMC HIGH
GAIN ANTENNA
(BLOCK II ONLY)

PLSS
ANTENNA
(BLOCK II
ONLY)

ANTENNA LOCATIONS

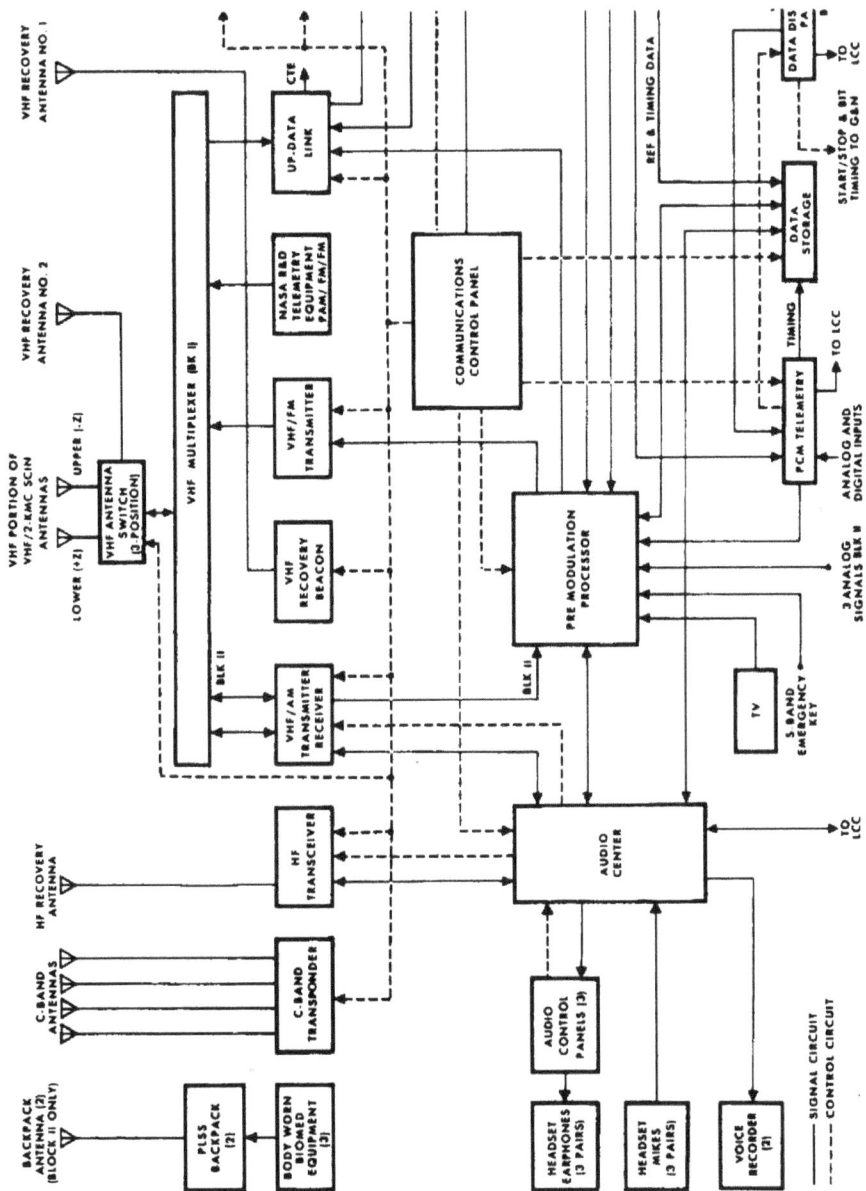

VHF RECOVERY ANTENNA NO. 1

VHF RECOVERY ANTENNA NO. 2

VHF PORTION OF VHF/2-KMC SCIN ANTENNAS

UPPER (-Z)

LOWER (+Z)

VHF ANTENNA SWITCH (3 POSITION)

VHF MULTIPLEXER (BK I)

UP-DATA LINK

CTE

NASA R&D TELEMETRY EQUIPMENT (PAM/FM/FM)

VHF/FM TRANSMITTER

VHF RECOVERY BEACON

COMMUNICATIONS CONTROL PANEL

VHF/AM TRANSMITTER RECEIVER

BLK II

BLK II

PRE MODULATION PROCESSOR

TV

S BAND EMERGENCY KEY

HF TRANSCEIVER

HF RECOVERY ANTENNA

C-BAND ANTENNAS

C-BAND TRANSPONDER

BACKPACK ANTENNA (2) (BLOCK II ONLY)

PLSS BACKPACK (2)

BODY WORN BIOMED EQUIPMENT (3)

AUDIO CENTER

AUDIO CONTROL PANELS (3)

HEADSET EARPHONES (3 PAIRS)

HEADSET MIKES (3 PAIRS)

VOICE RECORDER (2)

TO LCC

TO LCC

REF & TIMING DATA

DATA DIS PA

TO LCC

START/STOP & BIT TIMING TO G&N

DATA STORAGE

TIMING

PCM TELEMETRY

TO LCC

ANALOG AND DIGITAL INPUTS

3 ANALOG SIGNALS BLK II

———— SIGNAL CIRCUIT

—————— CONTROL CIRCUIT

44

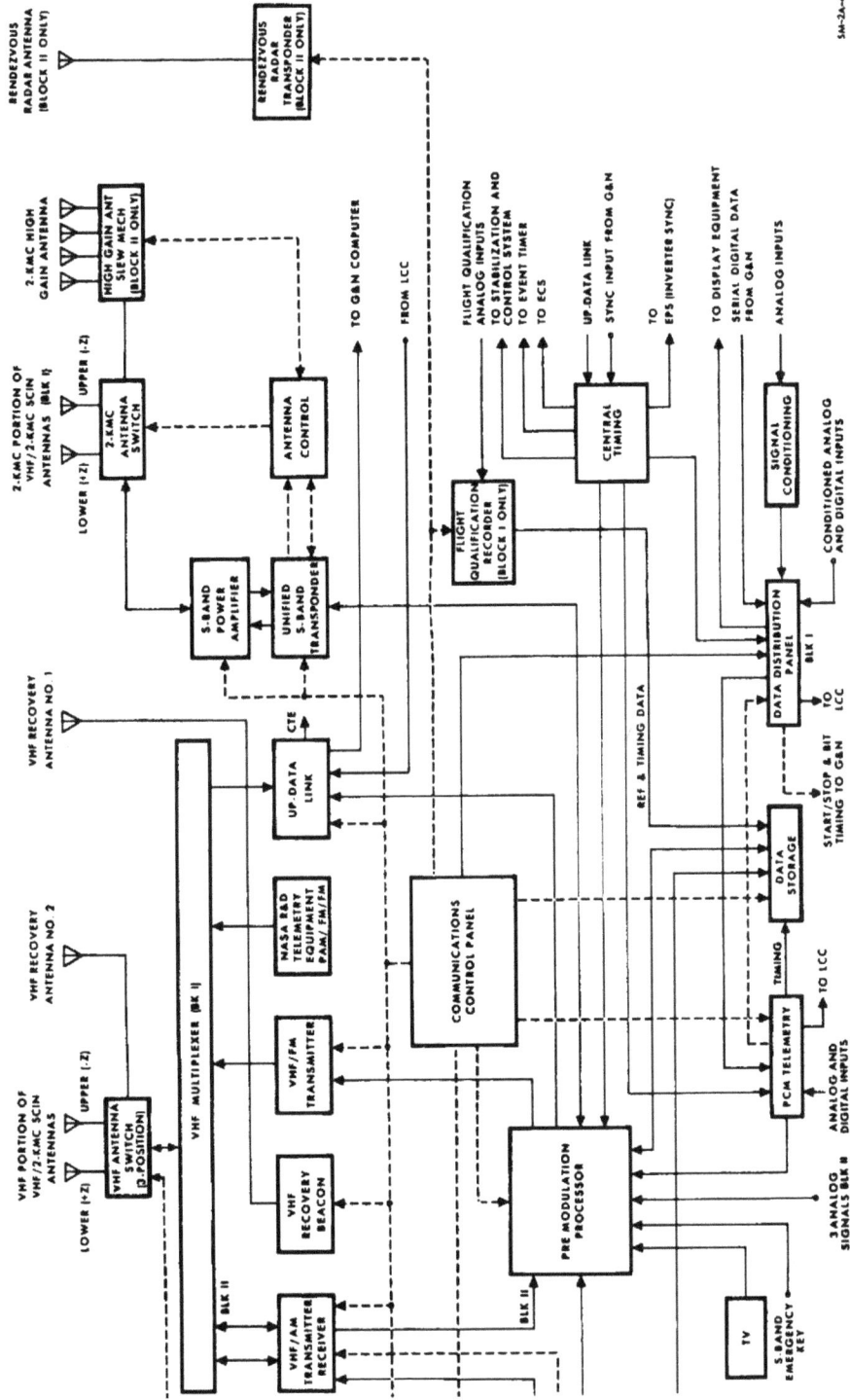

Figure 3-20. Telecommunications Systems - Antenna Location, Range, and Block Diagram

3-108. VOICE OPERATIONS.

3-109. S/C voice communications originate and terminate in the astronauts headsets. They are used for all voice transmission and reception, S/C intercommunications, and hardline communications with the launch control center (LCC) during prelaunch checkout. Each astronaut has an individual audio control panel, on the main display console (MDC), which enables him to select and control the inputs and outputs of his headset.

3-110. The headsets and audio control panels are connected to the audio center equipment which contains three identical audio center modules, one for each audio control panel and headset. The audio center equipment serves as the common assimilation and distribution point for all S/C audio signals. It is controlled by the audio control panels and the T/C controls on MDC panel No. 20. Depending on the mode of operation, the audio signals are routed to and from the applicable transmitter and receiver, the LCC, the recovery forces intercom, or the data storage equipment.

3-111. Three methods of voice transmission and reception are possible: the VHF/AM transmitter-receiver, the HF transceiver, or the S-band transmitter and receiver in the unified S-band equipment. Transmission is controlled by either the push-to-talk (PTT) switch, located in the electrical umbilical cord provided for each astronaut, or the voice-operated relay (VOX) circuitry. The PTT switch also serves as a keying switch during USBE emergency key transmission.

3-112. The VHF/AM transmitter-receiver equipment is used for voice communications with the MSFN during launch, ascent, and near-earth phases of the mission. The USBE is used during deep space phases of the mission when the S/C is not within VHF range of a ground station. When communications with the MSFN are not possible, limited capability exists to store audio signals on tape in the DSE for later transmission or playback on the ground after the mission is completed. For recovery operations during the post-landing phase of the mission, voice communications with the MSFN and recovery personnel are conducted over the VHF/AM transmitter-receiver equipment, the HF transceiver equipment, or the recovery forces intercom via the swimmers umbilical connector in the C/M forward compartment.

3-113. DATA OPERATIONS.

3-114. The S/C structure and operational systems are instrumented with sensors and transducers which gather data on their physical status. Biomedical data from sensors worn by the astronauts, TV data from the TV camera, and timing data from the central timing equipment are also acquired. These various forms of raw data are assimilated by the system, processed, and then transmitted to the MSFN. Data from the operational instrumentation may be stored in the DSE for later transmission or analysis. Analog data from the flight qualification instrumentation is stored in the flight qualification recorder for postflight analysis only.

3-115. Unconditioned analog and on-off event signals from instrumentation sensors are fed into the signal conditioning equipment where they are conditioned to a standard 5-volt d-c level. These signals are then sent to the data distribution panel which routes them to the pulse code modulation telemetry equipment and C/M displays. The PCM telemetry equipment combines the inputs from the SCE with other low-level analog inputs and converts them to a single, digital, modulating signal which is then routed to the premodulation processor equipment.

3-116. The PMP is the common assimilation, integration, and distribution center for nearly all forms of S/C data and provides the necessary interface with the RF equipment. In addition to the input from the PCM telemetry equipment, the PMP accepts recorded PCM and analog data from the DSE, video signals from the TV equipment, timing signals from the CTE, and audio signals from the audio center equipment when audio is transmitted by the USBE. These signals are modulated, mixed, and switched to the applicable transmitter or the DSE according to the mode of operation. Voice signals and up-data commands received over the USBE receiver from the MSFN are also supplied to the PMP which routes them to the audio center equipment and up-data link equipment, respectively. The UDL also contains its own receiver which is normally used during near-earth phases of the mission. S-band reception of up-data commands is used in deep space.

3-117. PCM/TLM data is transmitted to the MSFN by the VHF/FM transmitter equipment during near-earth phases of the mission. Transmission of TV or analog data is possible only by the USBE which is normally used in deep space but can also be used during launch and near-earth phases of the mission when within sight of an S-band equipped ground station. If the USBE is used, PCM/TLM data and voice signals can be transmitted with TV or analog data over the S-band link.

3-118. TRACKING AND RANGING OPERATIONS.

3-119. The T/C tracking and ranging equipment assists the MSFN in accurately determining the angular position and range of the S/C. Two methods are used: C-band tracking and S-band tracking.

3-120. C-band tracking is accomplished by the C-band transponder equipment and is used during all near-earth phases of the mission. It operates in conjunction with earth-based radar equipment. The C-band transponder transmits an amplified RF pulse in response to a properly coded, pulsed interrogation from the radar equipment. The range and accuracy of this equipment is thereby greatly extended over what would be possible by using skin-tracking techniques.

3-121. S-band tracking utilizing the USBE transponder is used in deep space. It operates in conjunction with MSFN equipment by providing responses to properly coded interrogations from the earth. When the USBE is in a ranging mode, the USBE transponder will receive PRN ranging code signals from the MSFN and respond by transmitting a similar signal. This method is used initially to establish an accurate measure of range to the S/C and subsequently at periodic intervals to update the doppler ranging data at the MSFN. The doppler measurements are obtained from the S-band carrier.

3-122. A VHF recovery beacon and an HF transceiver is provided to aid in locating the S/C during the recovery phase of the mission. The VHF recovery beacon equipment provides for line-of-sight direction finding capabilities by emitting a 2-second, modulated, RF transmission every 5 seconds. The HF transceiver equipment can be operated in a beacon mode to provide for beyond-line-of-sight direction finding by emitting a continuous wave signal.

3-123. INSTRUMENTATION SYSTEM.

3-124. The instrumentation system consists of those means required for the collection of data and is comprised of: operational, special, flight qualification, and scientific instrumentation. Equipment requirements include a variety of sensors, transducers, and photographic equipment which will be qualified prior to manned flight. Sensors and transducers are used for converting physical and electrical measurements into electrical signals. These signals are conditioned (by signal conditioners) to proper values for distribution to data utilization equipment and S/C displays.

3-125. Sensors and transducers, located strategically throughout the S/C, are positioned on the structure, within the operational systems, and for biomedical purposes are attached to the astronauts. Data may be transmitted to MSFN by way of the telecommunications system, displayed to the astronauts, or stored for evaluation at the completion of the mission.

3-126. The photographic equipment carried aboard the command module is provided for still photo and moving picture coverage.

3-127. OPERATIONAL INSTRUMENTATION.

3-128. Operational instrumentation consists of approximately 24 classes of transducers and is based on the following specific measurements:

Pressure	Attitude	Angular Position	Voltage
Temperature	Rates	Current	Frequency
Flow	Events	Quantity	RF Power

3-129. SPECIAL INSTRUMENTATION.

3-130. Special instrumentation consists of the equipment required for the checkout and monitoring of proton radiation detection and gas chromatograph. Requirements for additional equipment such as photographic, biomedical, fire-detection, and anthropomorphic dummies will vary, depending on the mission.

3-131. SCIENTIFIC INSTRUMENTATION.

3-132. Scientific instrumentation consists of the equipment required for various scientific experiments. The photographic equipment included in this group consists of a 35-mm still-photo camera and a 16-mm movie camera. Also included are data recorders which are installed when required.

3-133. FLIGHT QUALIFICATION.

3-134. Flight qualification consists of evaluation tests to ensure that the systems will furnish the instrumentation system with the required information in the specified form, from the source to the utilization point. The systems to be tested include sensors, associated equipment, subsystems, and special instrumentation such as optical, scientific, and biomedical. This data will be recorded on the flight qualification recorder for post-flight analysis.

3-135. CAUTION AND WARNING SYSTEM.

3-136. The caution and warning system (C&WS) monitors critical parameters of most S/C systems. Each malfunction or out-of-tolerance condition is brought to the attention of the crew by visual and aural means. The crew acknowledges the condition and resets the system for subsequent malfunction alerting.

3-137. C&WS OPERATION.

3-138. Malfunctions or out-of-tolerance conditions are sent to the C&WS as analog and discrete event inputs. This results in illumination of status lights that identify the abnormal condition, and activation of the master alarm circuit. Master alarm lights on the panels and an audio alarm tone in the headsets serve to alert the crew to each abnormal condition. Crew acknowledgement of the condition consists of resetting the master alarm circuit, thereby placing it in readiness should other malfunctions occur. C&WS operational modes are selected by the crew to meet varying conditions during the mission. The system also contains its own failure sensing signal.

3-139. CONTROLS AND DISPLAYS.

3-140. The operational spacecraft systems, located in the C/M and in the S/M, interface in varying degrees with control and display panels in the C/M cabin. This interface is provided to the extent necessary for the astronauts to adequately control and monitor the functions of the various systems. The controls and displays for most of the S/C systems are located on the main display console situated above the couches. (See figure 3-21.) This location permits frequent attention and quick control by the astronauts. Several of the S/C systems also have additional controls and displays elsewhere in the C/M cabin, as shown in figure 3-22. The majority of guidance and navigation system controls and displays are located on panels in the lower equipment bay adjacent to the G&N telescope and sextant. Those manual controls of the environmental control system that do not require frequent or time-critical actuation, are located in the left-hand equipment bay and the left-hand forward equipment bay. All of the controls in the waste management segment of the crew system are located on a panel in the right-hand equipment bay.

MAIN DISPLAY CONSOLE

A0118   SM-2A-567C

Figure 3-21. Controls and Displays - Main Display Console (Sheet 1 of 3)

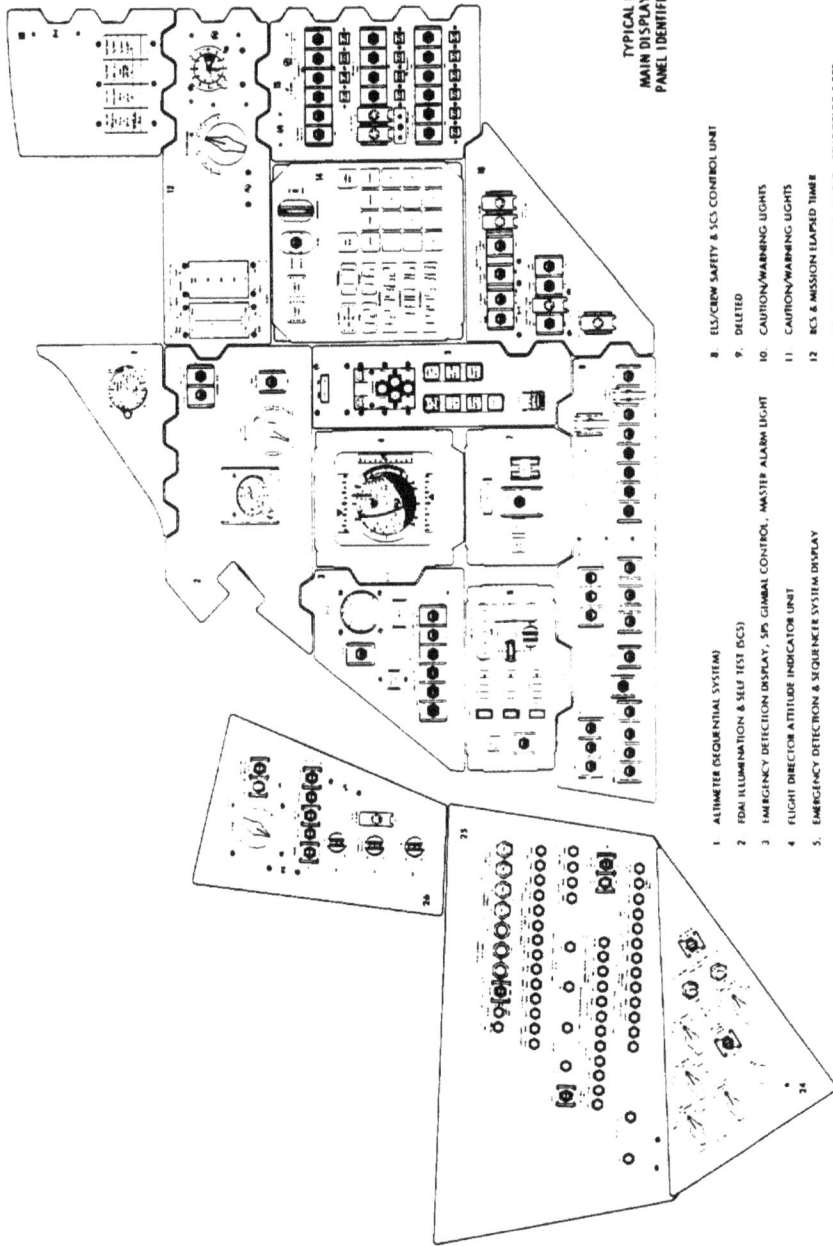

TYPICAL BLOCK I
MAIN DISPLAY CONSOLE—
PANEL IDENTIFICATION LIST

1. ALTIMETER (SEQUENTIAL SYSTEM)
2. FDAI ILLUMINATION & SELF TEST (SCS)
3. EMERGENCY DETECTION DISPLAY, SPS GIMBAL CONTROL, MASTER ALARM LIGHT
4. FLIGHT DIRECTOR ATTITUDE INDICATOR UNIT
5. EMERGENCY DETECTION & SEQUENCER SYSTEM DISPLAY
6. ATTITUDE SET & GIMBAL POSITION DISPLAY UNIT
7. ΔV DISPLAY UNIT

8. ELS/CREW SAFETY & SCS CONTROL UNIT
9. DELETED
10. CAUTION/WARNING LIGHTS
11. CAUTION/WARNING LIGHTS
12. RCS & MISSION ELAPSED TIMER
13. CAUTION & WARNING, AUDIO, CRYOGENIC & ECS
14. G&N COMPUTER CONTROL UNIT

3-53

3-53.a

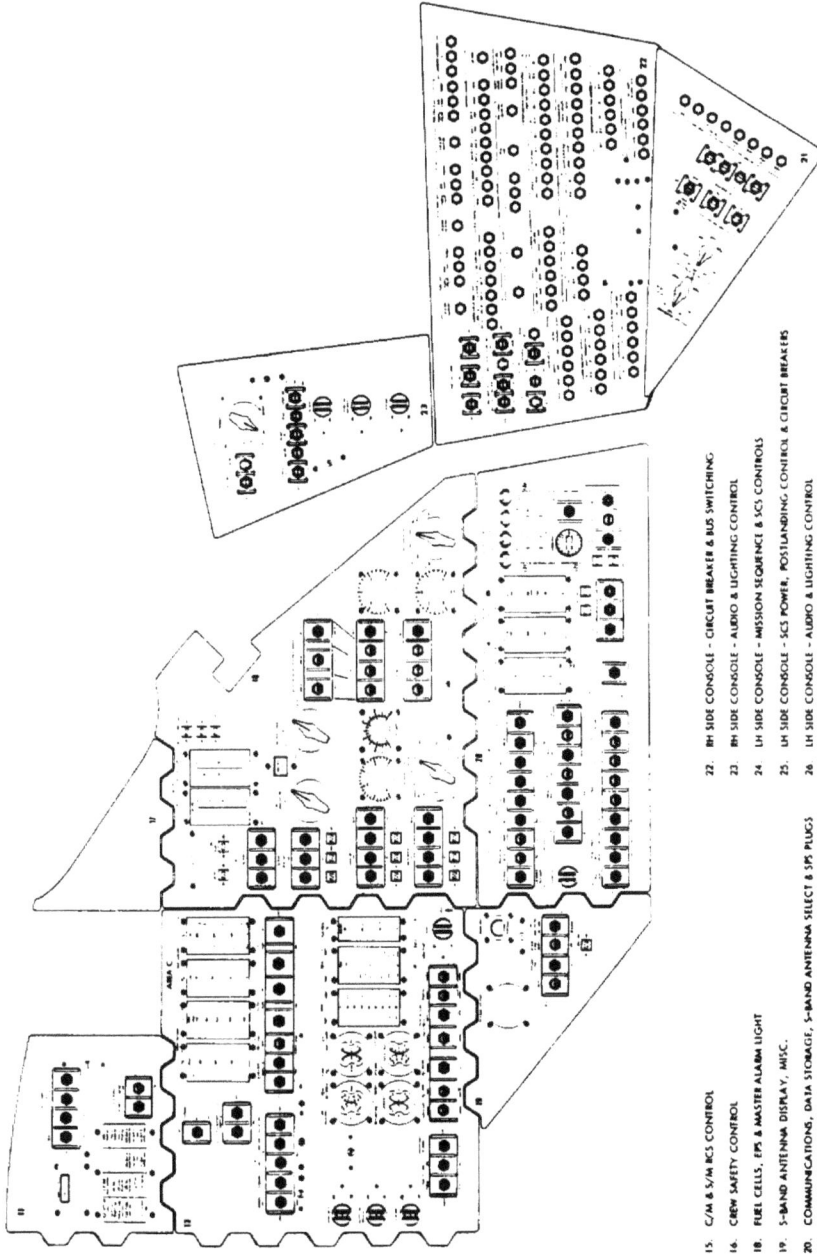

Figure 3-21. Controls and Displays - Main Display Console (Sheet 2 of 3)

15. C/M & S/M RCS CONTROL
16. CREW SAFETY CONTROL
18. FUEL CELLS, EPS & MASTER ALARM LIGHT
19. S-BAND ANTENNA DISPLAY, MISC.
20. COMMUNICATIONS, DATA STORAGE, S-BAND ANTENNA SELECT & SPS PLUGS
21. RH SIDE CONSOLE - CIRCUIT BREAKER & BUS SWITCHING

22. RH SIDE CONSOLE - CIRCUIT BREAKER & BUS SWITCHING
23. RH SIDE CONSOLE - AUDIO & LIGHTING CONTROL
24. LH SIDE CONSOLE - MISSION SEQUENCE & SCS CONTROLS
25. LH SIDE CONSOLE - SCS POWER, POSTLANDING CONTROL & CIRCUIT BREAKERS
26. LH SIDE CONSOLE - AUDIO & LIGHTING CONTROL

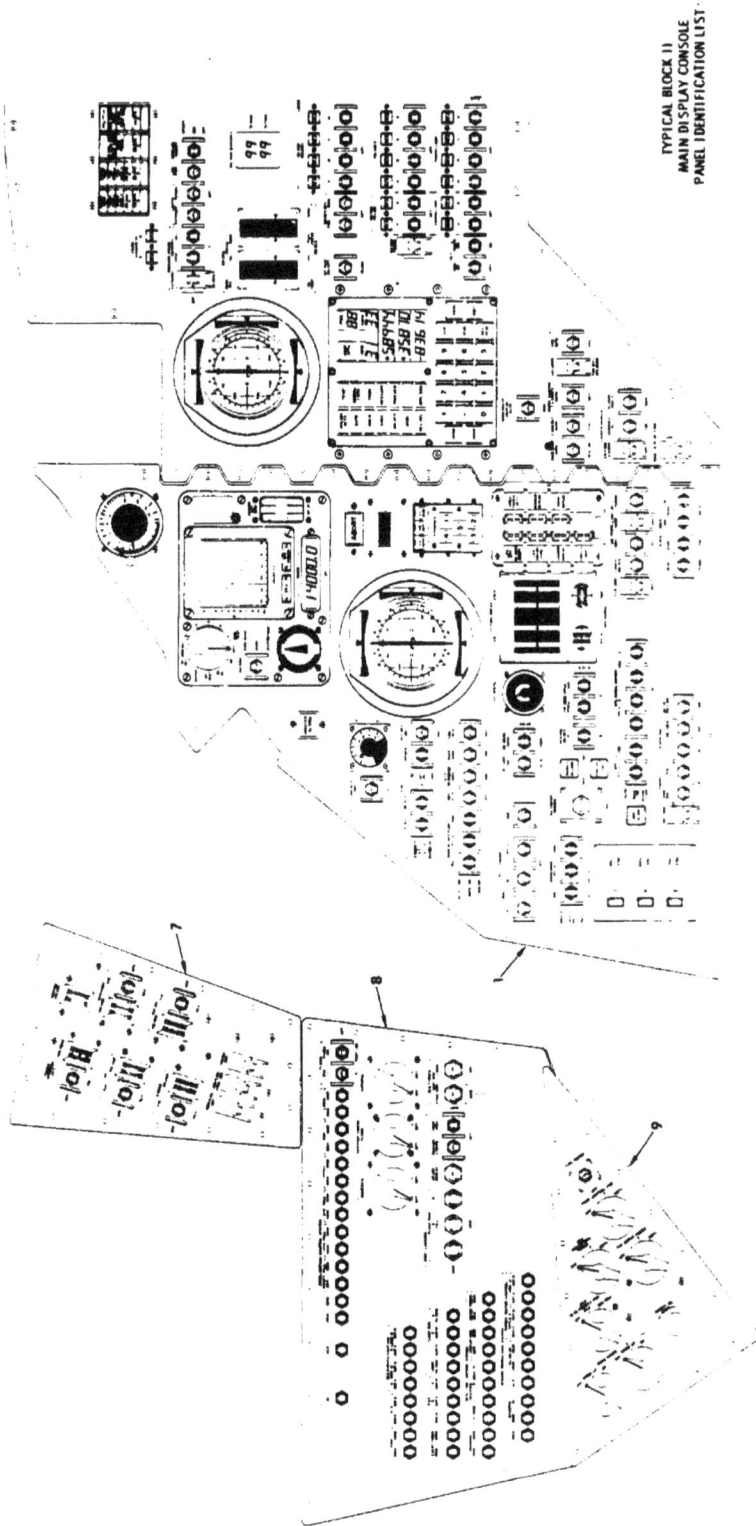

1 ALTIMETER, EMS AND ΔV REMAINING INDICATOR, FDAI, L/V FUEL AND ENGINES, SEQUENCER SYSTEM CONTROLS AND DISPLAYS, SPS GIMBAL CONTROL, ATTITUDE SET, G AND C, MASTER ALARM LIGHT, AND ABORT LIGHT.

2 CAUTION/WARNING LIGHTS, FDAI, G AND C DSKY, SEQUENCER SYSTEM, DOCKING SYSTEM, S/M AND C/M RCS, PHASE AND MISSION ELAPSED TIMERS, CRYOGENIC SYSTEM, ECS, AND HIGH GAIN ANTENNA.

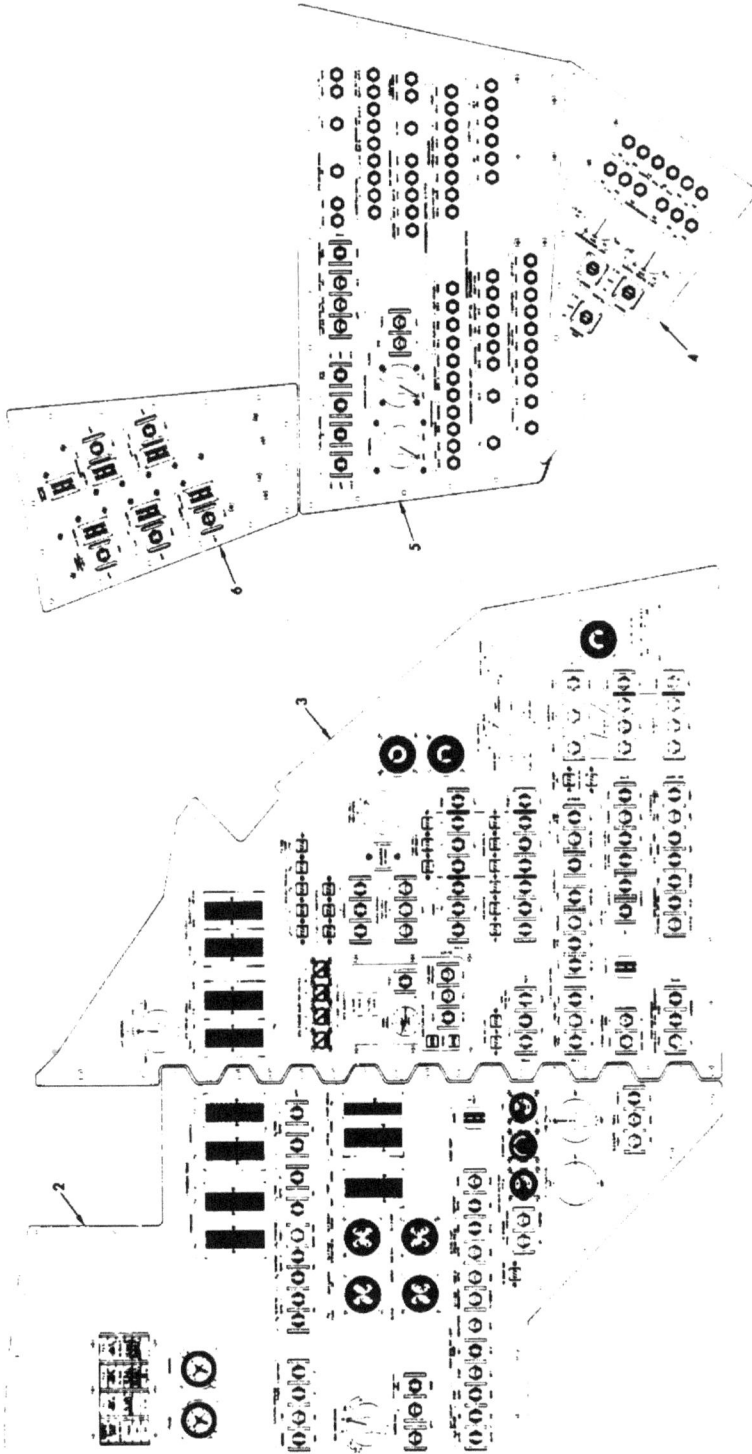

3 SPS, FUEL CELLS, EPS, I/C, AND MASTER ALARM LIGHT.

4 RH SIDE CONSOLE–SPS, I/C, AND ECS, BUS SWITCHING AND ECS CIRCUIT BREAKERS.

5 RH SIDE CONSOLE–BUS SWITCHING, CIRCUIT BREAKERS, AND INTERIOR LIGHT CONTROLS.

6 RH SIDE CONSOLE–I/C CONTROLS.

7 LH SIDE CONSOLE–EDS AND SCS CONTROLS.

8 LH SIDE CONSOLE–SCS, SEQUENCER SYSTEM, EDS, ECS, AND SPS CIRCUIT BREAKERS, POST-LANDING CIRCUIT BREAKERS AND SWITCHES, AND INTERIOR LIGHT CONTROLS.

9 LH SIDE CONSOLE–I/C CONTROLS.

Figure 3-21. Controls and Displays - Main Display Console (Sheet 3 of 3)

SM-2A-800A

3-56

3-54

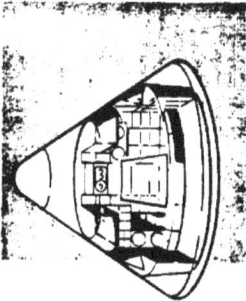

SM-2A-4750

3-57/3-58

3- 5 7-0.

EQUIPMENT IDENTIFICATION LIST

LH FORWARD EQUIPMENT BAY

1. PRESSURE SUIT CONNECTOR (ECS)
2. CABIN HEAT EXCHANGER SHUTTER (ECS)
3. CABIN TEMPERATURE CONTROL PANEL (ECS)
4. POTABLE WATER SUPPLY PANEL (ECS)
5. GIMBAL CLOCK AND EVENT TIMER (G&N)

LH EQUIPMENT BAY

6. OXYGEN CONTROL PANEL (ECS)
7. ENVIRONMENTAL CONTROL SYSTEM PACKAGE (BEHIND PANELS)
8. OXYGEN SURGE TANK (ECS)
9. CABIN PRESSURE RELIEF VALVE CONTROLS (ECS)

LOWER EQUIPMENT BAY

10. CONTROL PANEL (G&N)
11. G&N OPTICS
12. DATA STORAGE EQUIPMENT
13. COMMUNICATIONS MODULES
14. APOLLO GUIDANCE COMPUTER (G&N)
15. POWER SERVO ASSEMBLY (G&N)
16. BATTERY AND PYRO CONTROLS
17. CO2 ABSORBER CARTRIDGE STORAGE (ECS)
18. SCS MODULES
19. RATE AND ATTITUDE G-NO ASSEMBLIES (SCS)

RH FORWARD EQUIPMENT BAY

20. SYSTEMS MANAGEMENT

RH EQUIPMENT BAY

21. MISSION SEQUENCER AND SCIENTIFIC EQUIPMENT (BEHIND PANELS)
22. VACUUM CLEANER STORAGE
23. WASTE MANAGEMENT CONTROL PANEL
24. BUS TIE AND INVERTER POWER CONTROLS

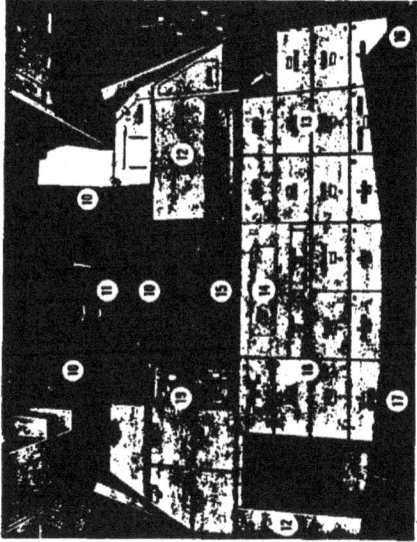

Figure 3-22. Command Module Equipment and Storage Bays (Block I)

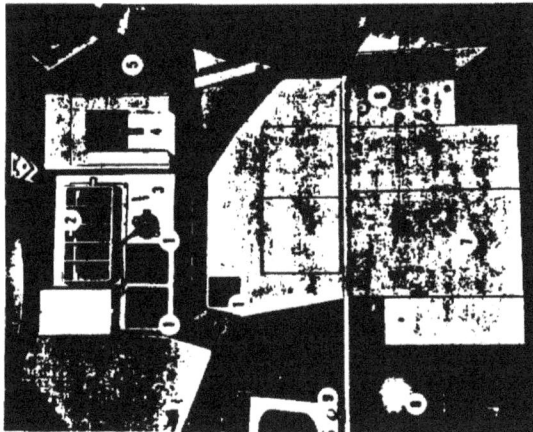

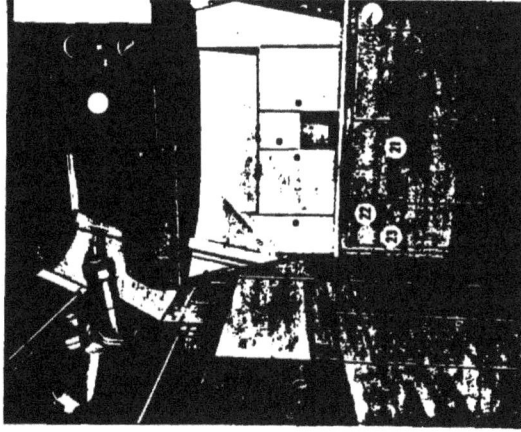

3-57

3-141. The displays that read out within given parameters are range-marked to aid the crew in more rapidly determining slight out-of-tolerance conditions. This is in addition to the caution/warning lights that will call the crew's attention to out-of-tolerance conditions. Regardless of the various types of controls in the C/M, provisions have been made for all controls to be operated with an astronaut's gloved hand.

3-142. DOCKING SYSTEM (BLOCK II).

3-143. The purpose of the docking system (figure 3-23) is to provide a means of connecting and disconnecting the LM/CSM during a mission and to provide a passageway to and from the LM when it is connected to the CSM. The docking system consists of the primary structure and the necessary hardware required to support the complete docking function between the CSM and LM.

3-144. The primary structure consists of the tunnel required for LM ingress and egress. The system contains hatches, latches, probe, drogue, illumination, and required sealing surfaces. The docking structure will be able to withstand load imposition from the docking maneuvers and from the modal characteristics that will exist between the LM and CSM.

3-145. Once the probe is engaged in the drogue, the probe attenuation system is activated to pull the LM against the CSM. This operation will lock the C/M to the LM with four semi-automatic latches, of the 12 latches around the forward circumference of the forward tunnel, providing an airtight seal. After the LM and CSM are latched together and sealed (initial docking), the following operations are accomplished: the pressure is equalized between the LM and C/M, the C/M forward tunnel hatch is removed, the eight remaining

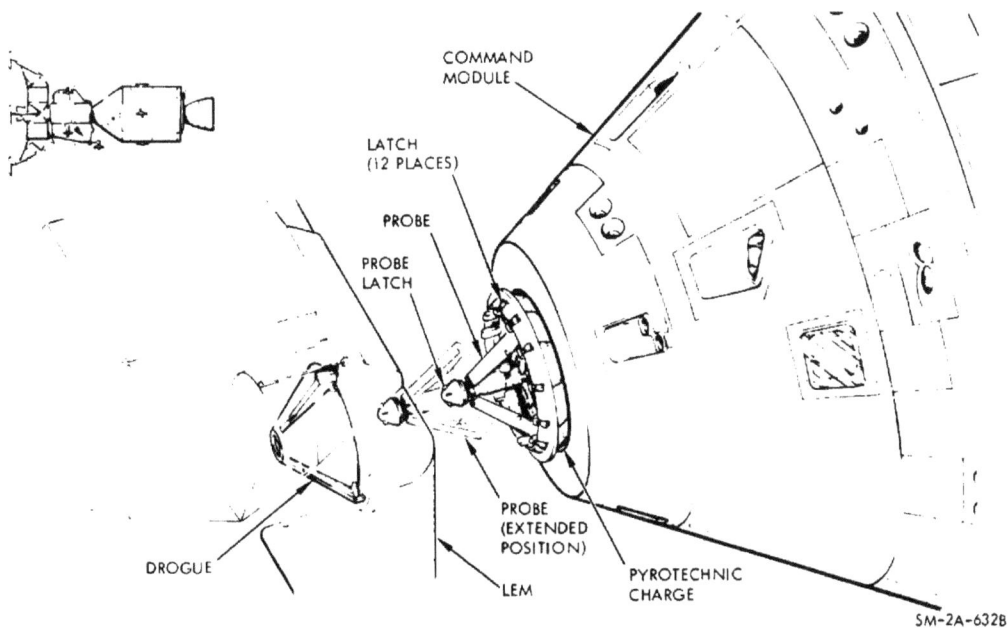

SM-2A-632B

Figure 3-23. Docking System (Block II)

manual and four semi-automatic latches are locked in place (final docking), an electrical umbilical is connected to the LM from the C/M, the probe and drogue mechanism is removed from the center of the tunnel, and the LM hatch is removed opening a passageway between the C/M and LM which allows transfer of equipment and crewmembers. In case a malfunction occurs which prevents docking or removal of the probe and drogue mechanism, provision has been made for emergency vehicular transfer (EVT). EVT involves transferring from one vehicle to another by way of the side hatches.

3-146. To release the LM from the CSM for lunar landing operations, the 12 manual latches are released, the hatches, probe, and drogue mechanisms are installed in their respective places, and the probe attenuation system is electrically released allowing the LM to separate from the CSM.

3-147. To release the LM from the CSM in preparation for the return trip to earth, all equipment with no further use to the astronauts is placed in the LM, including the probe and the drogue. After transfer of equipment and crewmembers, the C/M forward tunnel hatch is installed in place, resealing the C/M crew compartment. The LM is then released from the C/M by firing a pyrotechnic charge which is located around the circumference of the docking assembly.

3-148. CREWMAN OPTICAL ALIGNMENT SIGHT (BLOCK II).

3-149. The crewman optical alignment sight (COAS) is a sighting aid required for docking maneuvers after transposition of the CSM to assist the astronauts in accurately aligning the CSM with the LM. The LM also has an alignment sight that will be used in a similar manner after rendezvous is accomplished in lunar orbit. The COAS is a collimator-type optical device and provides the astronaut with a fixed line-of-sight attitude reference image. When viewed through the rendezvous window, the reference image appears to be the same distance away from the C/M as the LM target. It also has an elevation scale adjacent to the reference image, or reticle, with a range of +30° and -10°.

3-150. The COAS can be mounted on the inboard side of either the LH or RH rendezvous windows, and is designed to project an optical collimated alignment image. A light source with adjustable brightness control allows for proper brightness against all exterior background lighting conditions.

3-151. When the LM is the active docking vehicle, the COAS can be used as a backup to check the alignment of the LM with the CSM. The COAS can also be used as a replacement for the scanning telescope when necessary. Additional uses include: manual delta velocity reference, aiming cameras, and spacecraft and manual entry reference backup.

LUNAR MODULE

4-1. GENERAL.

4-2. This section contains basic data concerning the lunar module (LM). Information is given to indicate configuration, function, and interface of the various components in gross terms. For more detailed LM information, refer to the Lunar Module Familiarization Manual, LMA790-1.

4-3. The LM, illustrated in Figure 4-1, will carry two Apollo crewmembers from the orbiting CSM to the surface of the moon. The descent from lunar orbit will be powered by a gimbal-mounted rocket engine in the LM descent stage. As the LM nears the lunar surface, the descent engine will provide braking and hovering to allow lateral movement to a suitable landing area. The LM will provide a base of operations for lunar explorations. Food, water, electrical power, environmental control, and communications relay will sustain the crewmembers for a period up to 48 hours. One astronaut will explore the lunar surface while the other remains inside the LM. After 3 hours, the first astronaut will return and the second man will continue the exploration. Upon completion of exploration, the astronauts will prepare the LM ascent stage for ascent to intercept the orbiting CSM with power provided by the ascent rocket engine. The entire LM descent stage and other nonreturnable equipment is left on the moon. Upon rendezvousing and docking with the CSM, the crew will transfer from the LM to the CSM and, the LM will then be jettisoned and left as a lunar satellite.

4-4. LM CONFIGURATION.

4-5. STRUCTURE.

4-6. LM structural components are divided into ascent-stage and descent-stage structures. (See figure 4-2.) The ascent-stage structure consists of the following components:

- Crew compartment pressure shell
- Ascent tanks and engine support
- Equipment bay
- Equipment compartment
- Electronic replaceable assembly
- Oxygen, water, and helium tanks
- Reaction control system tanks and engine supports
- Windows, tunnels, drogue mechanism, and hatches
- Docking target recess
- Interstage fittings

The descent-stage structure consists of the following components:

- Descent engine tanks and engine support
- Landing gear assembly
- Secondary oxygen, water, and helium tanks

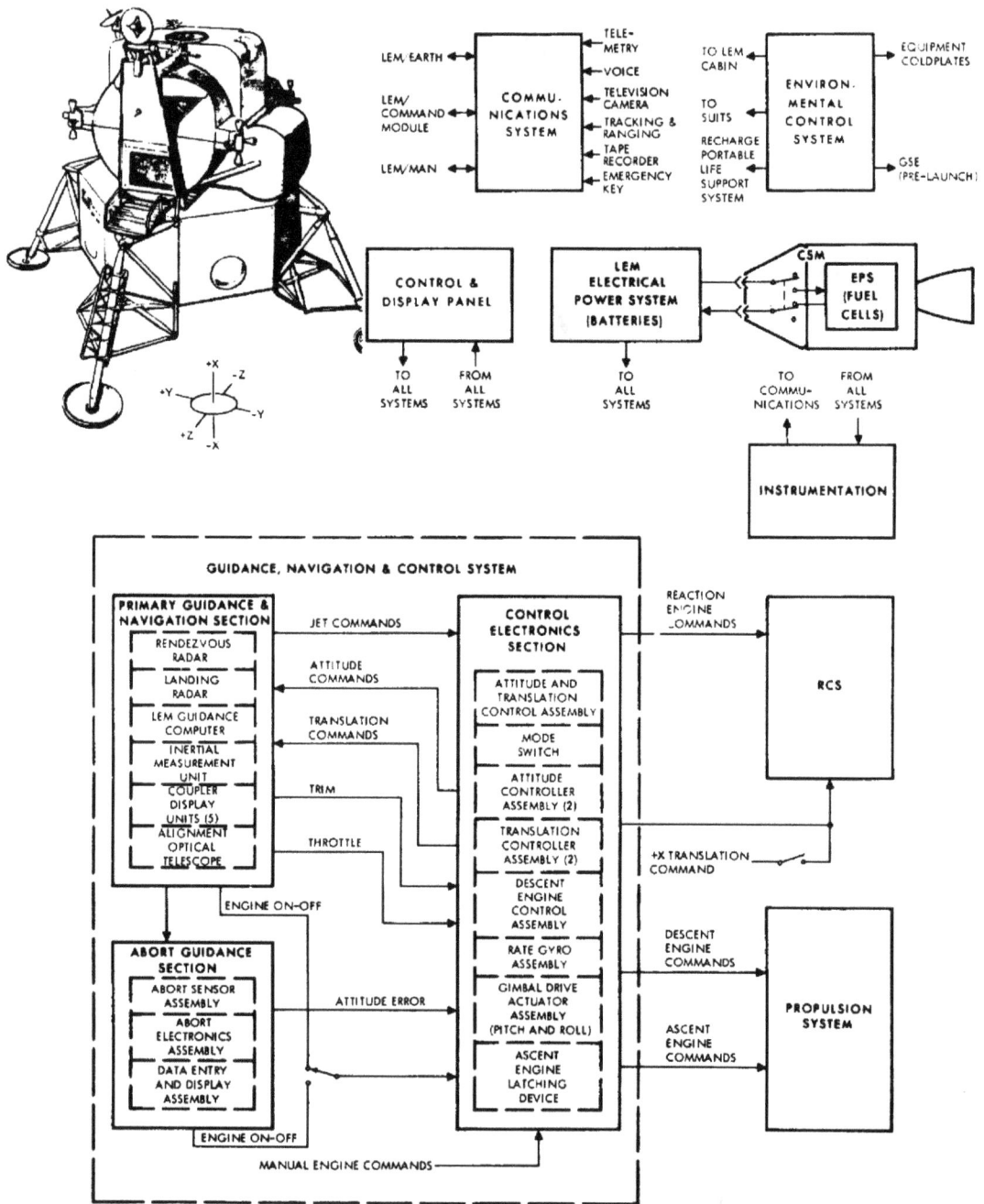

Figure 4-1. Lunar Module and Systems Block Diagram

SM-2A-494G

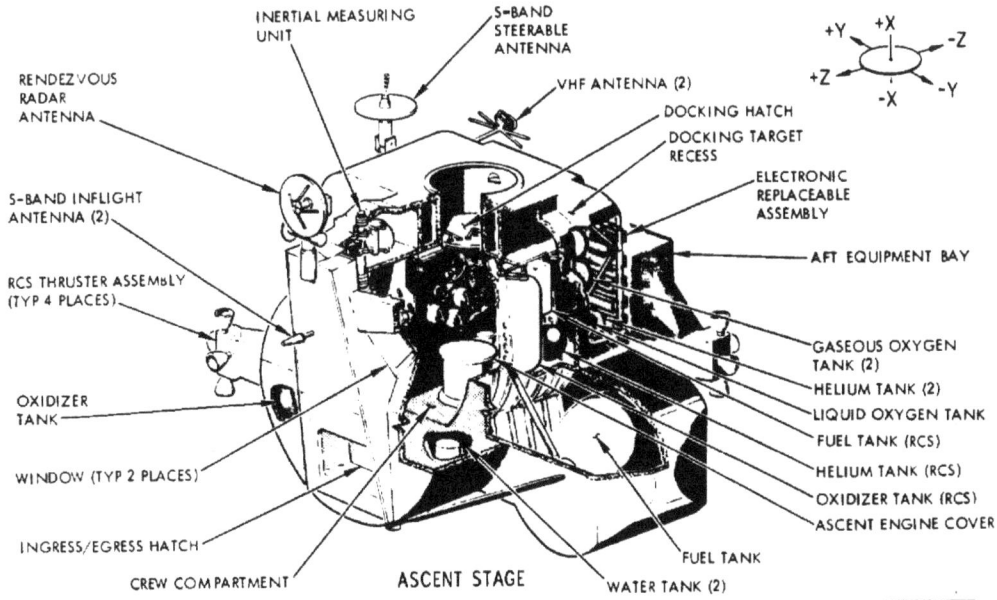

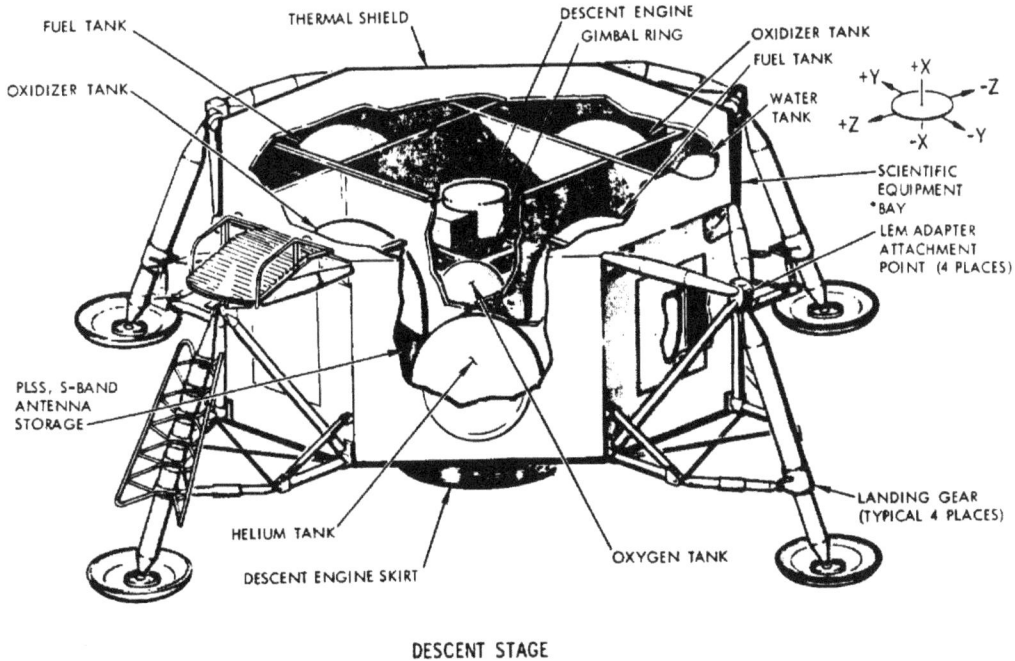

INERTIAL MEASURING UNIT

S-BAND STEERABLE ANTENNA

RENDEZVOUS RADAR ANTENNA

VHF ANTENNA (2)

DOCKING HATCH

DOCKING TARGET RECESS

ELECTRONIC REPLACEABLE ASSEMBLY

S-BAND INFLIGHT ANTENNA (2)

AFT EQUIPMENT BAY

RCS THRUSTER ASSEMBLY (TYP 4 PLACES)

GASEOUS OXYGEN TANK (2)

HELIUM TANK (2)

LIQUID OXYGEN TANK

FUEL TANK (RCS)

OXIDIZER TANK

HELIUM TANK (RCS)

OXIDIZER TANK (RCS)

WINDOW (TYP 2 PLACES)

ASCENT ENGINE COVER

INGRESS/EGRESS HATCH

FUEL TANK

CREW COMPARTMENT

ASCENT STAGE

WATER TANK (2)

+Y +X -Z

+Z -Y

-X

FUEL TANK

THERMAL SHIELD

DESCENT ENGINE GIMBAL RING

OXIDIZER TANK

OXIDIZER TANK

FUEL TANK

WATER TANK

+Y +X -Z

+Z -Y

-X

SCIENTIFIC EQUIPMENT BAY

LEM ADAPTER ATTACHMENT POINT (4 PLACES)

PLSS, S-BAND ANTENNA STORAGE

LANDING GEAR (TYPICAL 4 PLACES)

HELIUM TANK

DESCENT ENGINE SKIRT

OXYGEN TANK

DESCENT STAGE

SM-2A-599C

Figure 4-2. LM Ascent and Descent Stages

- Scientific equipment bay
- Interstage fittings
- Antenna storage bay
- Battery storage bay

4-7. LM OPERATION.

4-8. Operation of the LM is under crew control. Controls and displays provide monitoring and allow control of the various systems. Warning and caution lights are provided on two centrally located panels. A malfunction in any of the systems will light a specific indicator, denoting the malfunctioning system. General operation of the various systems (figure 4-1) is explained in paragraphs 4-9 through 4-29.

4-9. GUIDANCE, NAVIGATION, AND CONTROL SYSTEM.

4-10. The guidance, navigation and control system is an inertial system aided by optical sighting equipment and radar. The function of the guidance, navigation, and control system is to provide and/or maintain the following:

- Position data
- Velocity
- LM attitude
- Altitude
- Rate of ascent/descent
- Range (from command/service module)
- Range rate
- Control command

The system consists of three sections; a primary guidance and navigation section, a control electronics section, and an abort guidance section. A LM guidance computer (LGC), inertial measurement unit (IMU), coupling display units (CDU), optical telescope, rendezvous radar, and landing radar comprise the primary guidance and navigation section.

4-11. The guidance, navigation, and control system monitors and controls the attitude of the LM, provides guidance backup, enables automatic or manual control modes, and controls propulsion engine gimbals. Attitude error signals are generated in the primary guidance and navigation section and the abort guidance section, and routed to the control electronics section. The control electronics section provides attitude error correction signals and propulsion commands. An attitude and translation control assembly provides signal routing and mode control for the guidance, navigation, and control system. Mode control enables manual or automatic control of the LM attitudes through the reaction control system and LM velocity through the propulsion system. Rate gyros and an in-flight monitor provide control and display information. If the primary guidance and navigation section fails, the abort guidance section mode may be selected to take over the guidance functions. Propulsion engine, gimbal controls, and firing commands are controlled by the primary guidance and navigation section and control electronics section to ensure thrusting through the LM center of gravity. Backup control is provided by the abort guidance section.

4-12. RADAR SYSTEMS. There are two separate radar systems which aid the guidance, navigation, and control system. The rendezvous radar provides range and range-rate information, as well as azimuth (target bearing) information, utilizing a gimbal-mounted antenna. The landing radar provides altitude and altitude change-rate information utilizing a two-position antenna. The control and display panels provide crew control of the radar system and display of the information received. A storage buffer receives the acquired information from signal conditioners; then a high-speed counter, timer by the LGC, converts the information into representative digital form which is fed into the LGC.

4-13. PROPULSION SYSTEM.

4-14. Two rocket engines provide the power required for descent and ascent. The engines use pressure-fed liquid propellants. The propellants consist of a 50:50 mixture of UDMH and hydrazine as fuel, and nitrogen tetroxide as the oxidizer. Ignition is by hypergolic reaction when the fuel and oxidizer are combined. The descent engine, fuel tanks, oxidizer tanks, and associated components are located in the LM descent stage. Provision is made to throttle the descent engine to enable velocity control. Gimbal mounting of the engine provides hovering stability. The ascent engine is centrally mounted in the LM, and is of fixed-thrust, nonthrottling configuration, mounted in a fixed position. The propellant supply of the ascent engine is interconnected with the reaction control system propellant supply. Control of the engines may be either manual or automatic, with automatic control maintained by the LGC through the guidance, navigation, and control system.

4-15. REACTION CONTROL SYSTEM.

4-16. LM attitude control is provided by 16 small rocket engines mounted in four clusters. Each cluster consists of four engines mounted 90 degrees apart. The engines are supplied by two pressure-fed propellant systems. The propellants are the same as those used by the propulsion engines. The propellant supply to the reaction control system engines is also interconnected to the ascent engine propellant supply, allowing extended use of the reaction control engines. Reaction engine commands may be manual or automatic, and are applied through the guidance, navigation, and control system.

4-17. ENVIRONMENTAL CONTROL SYSTEM.

4-18. Environmental control is maintained inside the LM cabin. Portable life support systems, in the form of backpacks, supply a controlled environment in the pressure suits to allow exploration of the lunar surface. Oxygen, water, and water-glycol are used for environmental control. Pure oxygen is stored in a tank located in the ascent stage. The pure oxygen is conditioned for use by mixing it with filtered oxygen. The descent stage contains a tank which stores additional oxygen in the super-critical (liquid or extremely cold) state. Potable water for drinking, food preparation, and the backpacks, is stored in a water tank. Temperature control of the cabin and electronic equipment is provided by a water-glycol cooling system. The coolant is pumped through the electronic equipment coldplates and heat exchangers, and filtered. Cabin temperature control is monitored by temperature sensors and maintained by a temperature controller. The portable life support system (PLSS) provide necessary oxygen, water, electrical power, and a communications link to enable the LM crewmembers exploring the lunar surface to reamin in contact with each other, the CSM and MSFN. The backpacks can be used approximately 4 hours, after which the oxygen tank must be refilled and the batteries recharged from the environmental control system.

4-19. ELECTRICAL POWER SYSTEM.

4-20. Electrical power is provided by six silver oxide-zinc, 28-vdc batteries, four in the descent stage and two in the ascent stage. Two additional batteries are provided specifically for explosive devices. The batteries will supply sufficient power to maintain essential functions of the LM. Power distribution is provided by three buses; the commander bus, the system engineer bus, and the a-c bus. The commander and system engineer buses (28 vdc) supply power to components which must operate under all conditions. Power to all other components is provided by the a-c bus. The a-c bus is provided with 115-vac 400-cps power by one of two inverters selected by a crewmember. The two electroexplosive device batteries provide power to fire explosive devices for the landing gear uplock, stage separation, and helium pressurizing valves in the propulsion and reaction control systems.

4-21. COMMUNICATIONS.

4-22. Communications aboard the LM are divided into three systems, listed as follows:

- LM-earth system
- LM-command module system
- LM-crewmember system

The LM-earth system will provide telemetry, television, voice, taped playback, hand-key, and transponder communication to earth. Return from earth will be in the form of voice and digital up-data. The LM-C/M system will provide voice communications between the orbiting C/M and LM. PCM telemetry data at 1.6 kilobits per second can be transmitted from the LM to the command module. The LM-crewmember system provides inter-communication for the LM crew, and voice suit telemetry communication is provided by the backpacks when one crewmember is on the lunar surface conducting explorations.

4-23. INSTRUMENTATION.

4-24. Operational instrumentation senses physical data, monitors the LM subsystems during the unmanned and manned phases of the mission, prepares LM status data for transmission to earth, stores time-correlated voice data as required, and provides timing frequencies for the other LM subsystems. The instrumentation subsystem consists of sensors, signal conditioning electronics assembly, caution and warning electronics assembly, pulse code modulation and timing electronics assembly, and the data storage electronics assembly.

4-25. CONTROL AND DISPLAY PANELS.

4-26. The controls and display panels contain controls, monitoring instruments, and warning indicators to enable the crewmembers to maintain full knowledge of the status of various systems. Manual overrides allow the crewmembers to compensate for any deviations not allowed in automatic systems operation, or to take over a malfunctioning operation.

4-27. CREW PROVISIONS.

4-28. The crew provisions consist of miscellaneous equipment necessary to support two crewmen in the descent, the 24- to 48-hour exploration, and the ascent phases. The items included are listed as follows:

- Extravehicular mobility unit (Includes space suits, garments, and PLSS)
- Astronaut supports and restraints
- Lighting
- First-aid kit

- Food storage and water dispensing
- Waste management section
- Medical kit

4-29. SCIENTIFIC INSTRUMENTATION.

4-30. Scientific instrumentation will be carried to the lunar surface aboard the LM to enable the crewmembers to acquire samples and data concerning the lunar environment. A list of typical instrumentation to be used is as follows:

- Lunar atmosphere analyzer
- Gravitometer
- Magnetometer
- Penetrometer
- Radiation spectrometer
- Specimen return container
- Rock and soil analysis equipment
- Seismograph
- Soil temperature sensor
- Self-contained telemetering system
- Camera
- Telescope

As additional data concerning the lunar environment become available, this list will be altered.

APOLLO SPACECRAFT MANUFACTURING

5-1. GENERAL.

5-2. This section describes the manufacturing fabrication, assembly, subsystems installation, and functional checkout of Apollo spacecraft structures and systems. Manufacturing techniques and processes utilized, incorporate features applicable to the special requirements of the Apollo spacecraft. Final assembly, subsystems installation, and functional checkout is performed in environmentally controlled cleaning rooms, providing control of humidity, temperature, and sources of contamination.

5-3. SPACECRAFT MAJOR ASSEMBLIES.

5-4. The spacecraft is comprised of four major assemblies consisting of the launch escape system, the command module, service module, and spacecraft LM adapter (SLA). The SLA houses the lunar module.

5-5. LAUNCH ESCAPE SYSTEM STRUCTURE.

5-6. The launch escape system structure (figure 5-1) consists of a nose cone, canard assembly, pitch control motor, tower jettison motor, launch escape motor, skirt structure assembly, tower structure assembly, and hard and soft boost protective covers. The entire system is 33 feet long.

5-7. The canard assembly is made from Inconel nickel and stainless steel skins riveted together. The tower structure assembly is a fusion-welded, titanium tubing structure with fittings at each end for attachment to the skirt structure assembly and the

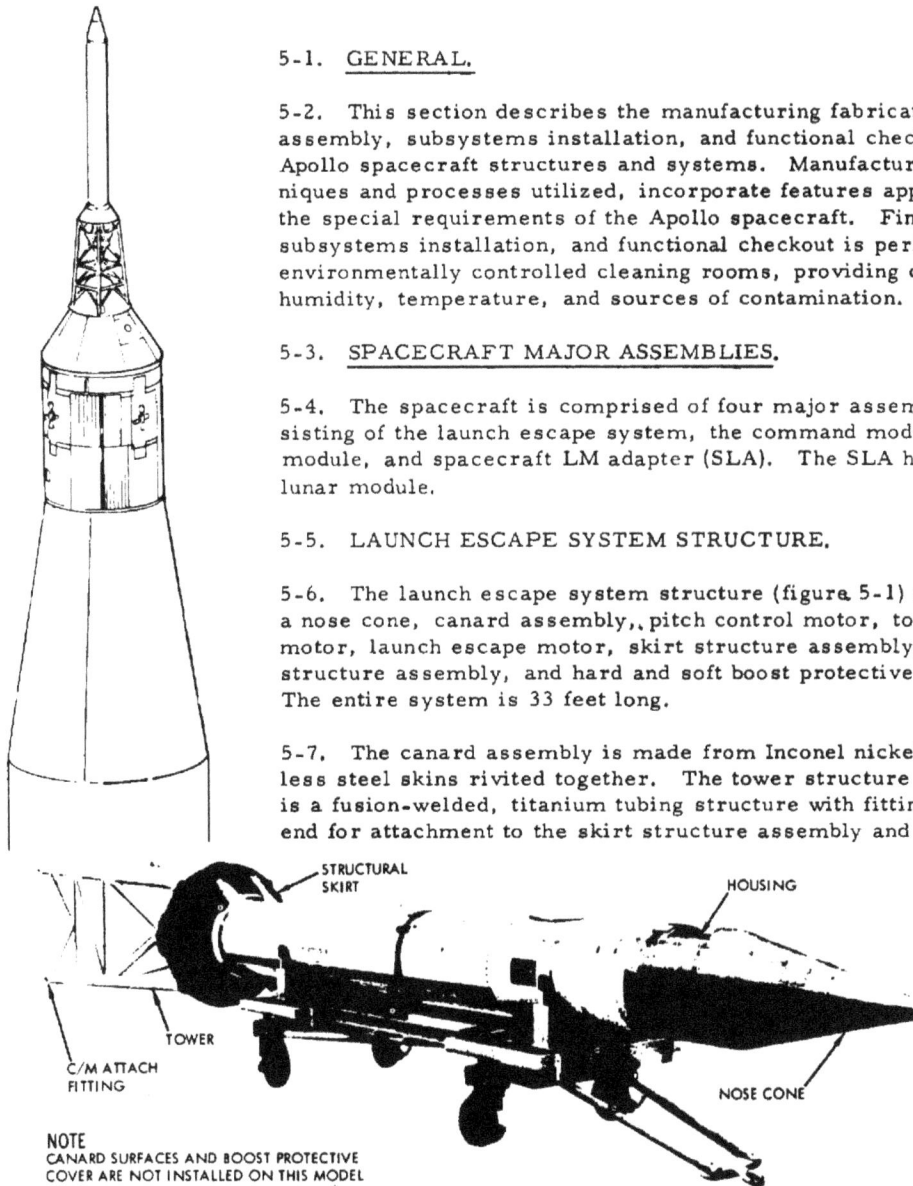

NOTE
CANARD SURFACES AND BOOST PROTECTIVE
COVER ARE NOT INSTALLED ON THIS MODEL

Figure 5-1. Launch Escape System Structure

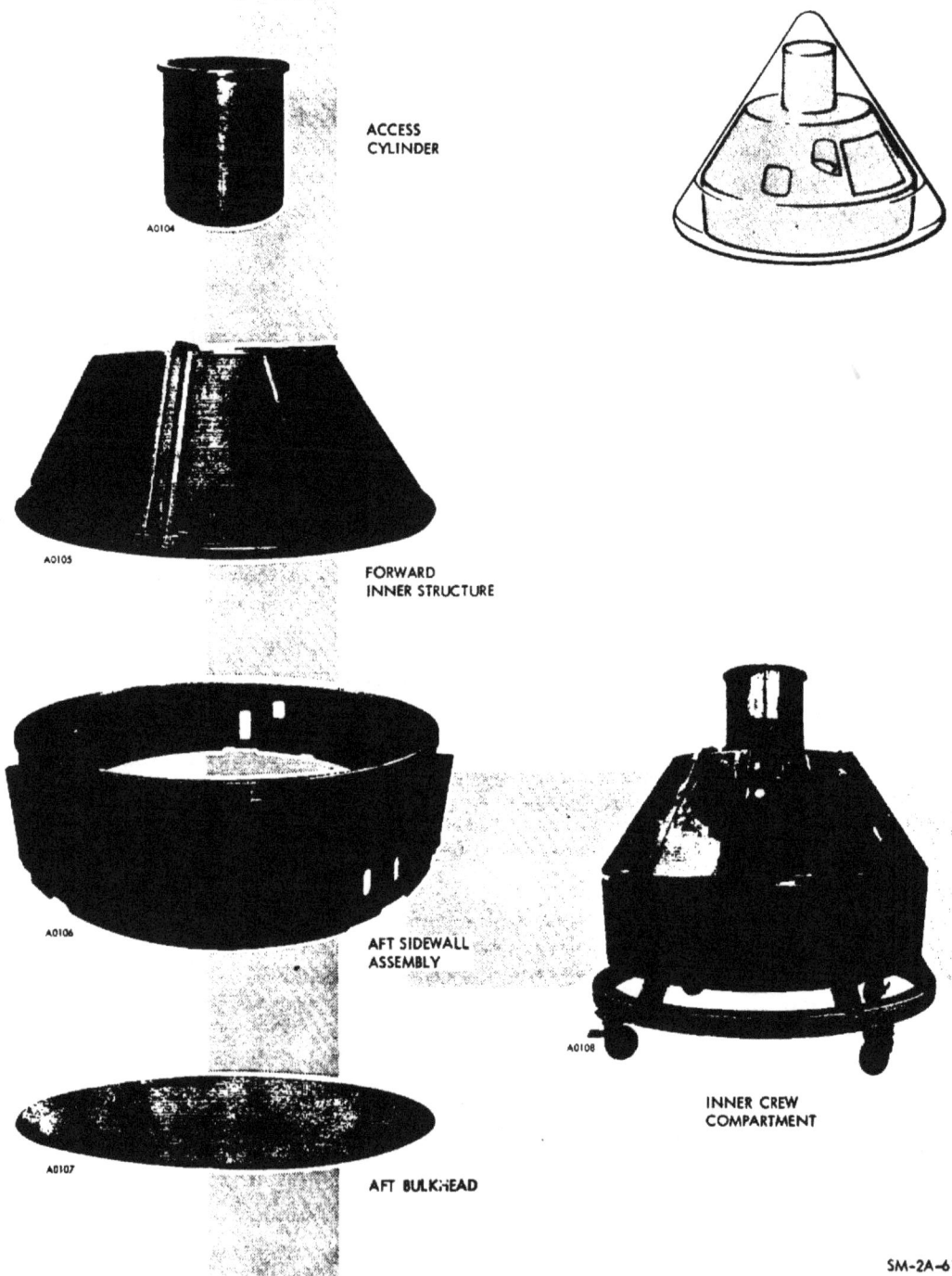

ACCESS
CYLINDER

A0104

FORWARD
INNER STRUCTURE

A0105

AFT SIDEWALL
ASSEMBLY

A0106

INNER CREW
COMPARTMENT

A0108

AFT BULKHEAD

A0107

SM-2A-610A

Figure 5-2. Command Module Inner Crew Compartment Structure

command module release mechanism. The ballast enclosure and pitch control motor structure assemblies are fabricated from nickel alloy steel sheet metal skins, which are riveted to ring bulkheads and frames. The skirt assembly is made from titanium and is welded and riveted during construction. The boost protective covers, constructed of glass cloth, phenolic honeycomb, and ablative cork, are fastened to the bottom of the tower.

5-8. COMMAND MODULE STRUCTURE.

5-9. The basic structure of the command module consists of a nonpressurized outer heat shield and a pressurized inner crew compartment. Bolted frame assemblies secure the heat shield to the aft compartment while I-beam stringers are used to mechanically fasten the inner and outer structures in the forward crew compartment area. A two-layer micro-quartz fiber insulation is installed between the inner and outer structures. Ablative materials are applied to the outer heat shield to protect the C/M against aerodynamic heating during entry into the atmosphere of the earth.

5-10. INNER CREW COMPARTMENT STRUCTURE. The inner crew compartment consists basically of a forward section containing an access cylinder welded to the forward bulkhead and cone, and an aft section containing the sidewall and bulkhead. (See figure 5-2.) These two sections are welded into subassemblies, then honeycomb bonded, trimmed, butt-fusion welded (figure 5-3), and the closeout area is filled with aluminum honeycomb material. Face sheets of aluminum alloy overlap the two sections and are bonded in place. Secondary structure equipment bays, housing the various subsystems and storage areas, are located within the inner crew compartment. Refer to section III for a description of the systems installed in the C/M.

5-11. HEAT SHIELD STRUCTURE. The heat shield structure (figure 5-4) of the command module consists of the apex cover, forward heat shield, crew compartment heat shield, and aft heat shield. The apex cover, used on Block I S/C, is replaced with the C/M portion of the docking system on Block II S/C.

5-12. The forward heat shield consists of four honeycomb panels, two rings, and four launch escape tower leg wells. The panels are placed in a jig singly and trimmed longitudinally. The tower leg wells are installed, trimmed, and welded. The panels are then installed in a jig which accommodates all four panels, trimmed longitudinally, and butt-fusion welded. The welded panels and the rings are placed in another jib for circumferential trim, and the rings are then welded to the top and bottom of the panels. The completed assembly is fit-checked to the crew compartment and aft heat shields, and then removed for the application of ablative material.

5-13. The crew compartment heat shield is formed from steel honeycomb panels and rings. The panels are joined together by machined edge-members which provide door-opening lands and are attached to the inner crew compartment structure. The panels and rings are installed in a series of jigs for assembly, trimming, and welding. The welded sections are then placed in a large fixture for precision machining of the top and bottom rings. The assembly is fit-checked with the inner crew compartment, forward heat shield, and aft heat shield, and then removed for the application of ablative material.

5-14. The aft heat shield assembly consists of four brazed honeycomb panels joined laterally by fusion welds and attached to a 360-degree machined ring by spot-welded sheet metal fairings, using conventional mechanical fasteners. Holes for the inner and outer C/M component attach points and the tension tie locations are cut through the assembly by (tre-pan) machining. The completed aft heat shield is fit-checked with the crew compartment heat shield prior to the application of ablative material.

Figure 5-3. Trim and Weld Closeout Operation

5-15. SERVICE MODULE STRUCTURE.

5-16. The service module consists basically of a forward and aft bulkhead, radial beams and outer panels, and is constructed primarily of aluminum alloy which is honeycomb bonded. The eight outer panels are aluminum honeycomb bonded between aluminum face sheets. The forward bulkheads of the Block II S/M will be constructed of sheet metal with the aft bulkhead remaining aluminum-bonded honeycomb. Six radial beams, which divide the cylinder into compartments, are machined and chem-milled to reduce weight in non-critical stress areas. Beams, bulkheads, and support shelves form the basic structure. (See figure 5-5.)

5-17. The fuel and oxidizer tanks, hydrogen and oxygen tanks, fuel cells, reaction control system, service propellant system, antenna equipment, electrical power system, and part of the environmental control system, are housed in the service module.

LES TOWER
LEG WELL
(4 PLACES)

FORWARD
HEAT SHIELD

APEX
COVER

A0112

CREW
COMPARTMENT
HEAT SHIELD

A0113

AFT HEAT SHIELD   A0114

SM-2A-613B

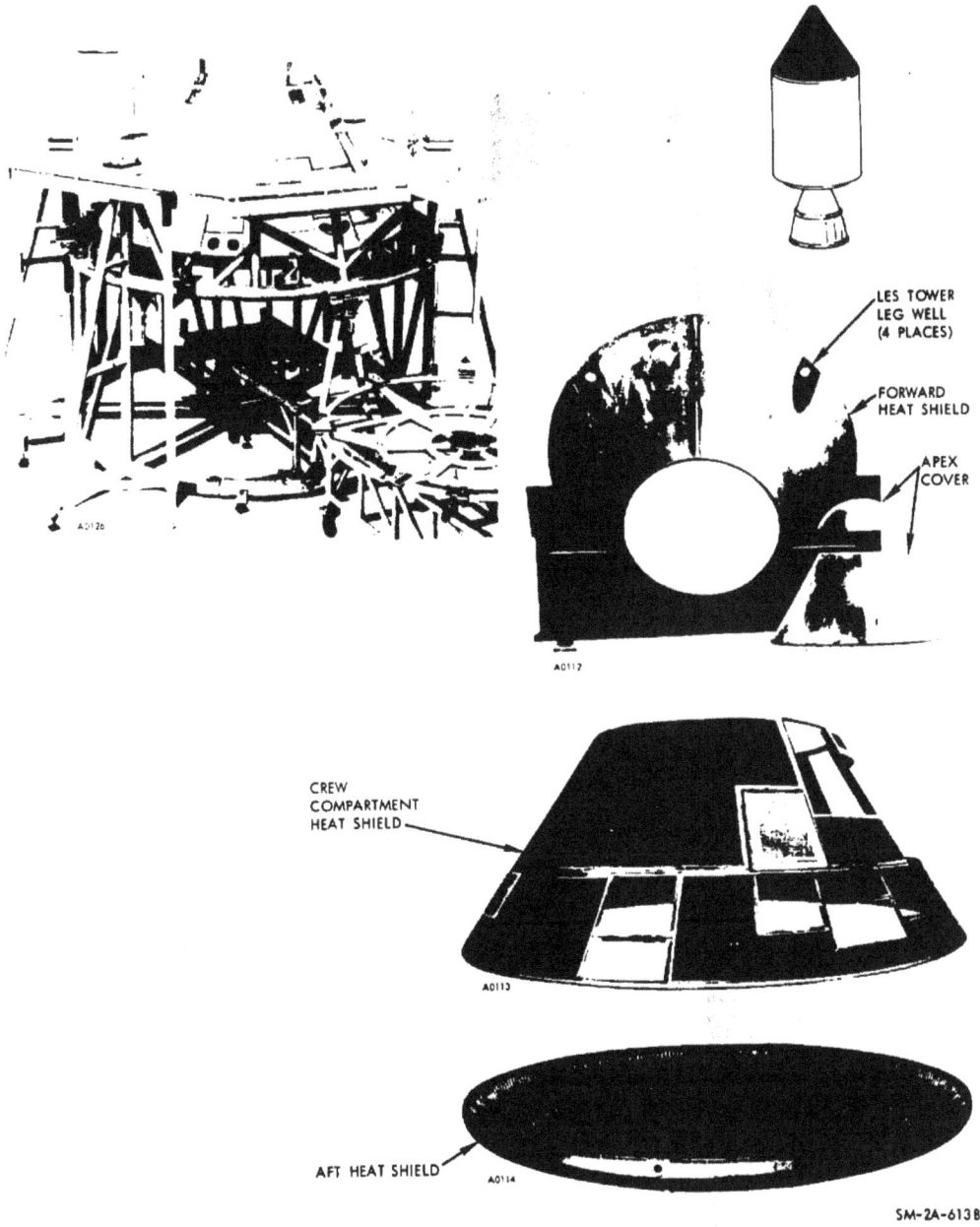

Figure 5-4. Command Module Heat Shield Structure

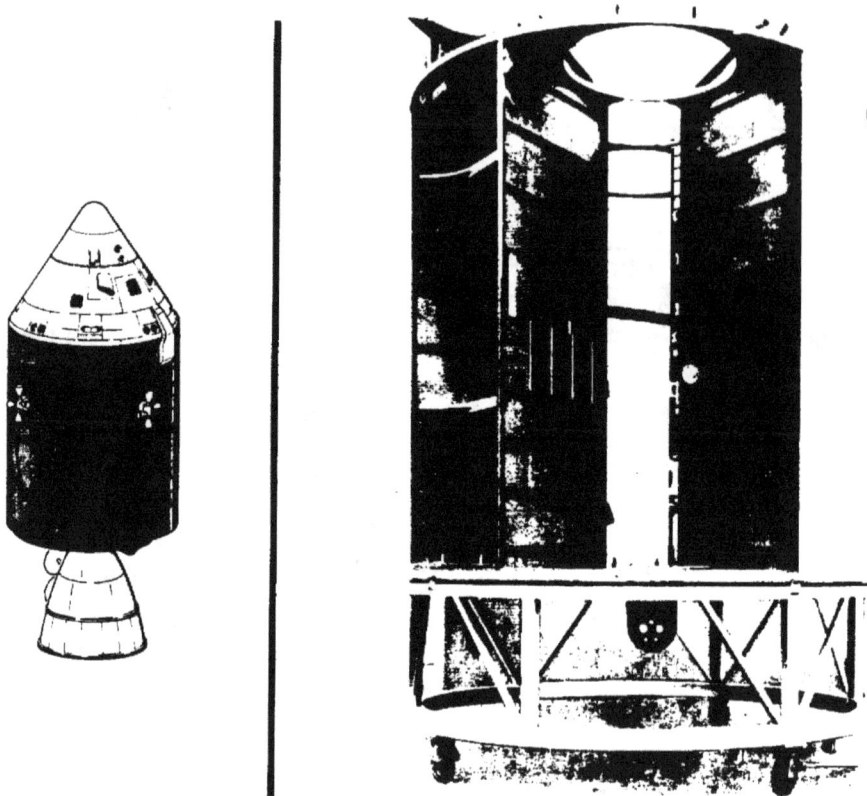

Figure 5-5. Service Module Structure

5-18. SPACECRAFT LM ADAPTER.

5-19. The spacecraft LM adapter (SLA) is a truncated cone, constructed of bonded aluminum honeycomb, which connects the S/M with the S-IVB instrument unit and houses the LM. The adapter is 28 feet in length, 12 feet 10 inches in diameter at the forward end, and 21 feet 8 inches in diameter at the aft end. The SLA consists of eight 2-inch-thick bonded aluminum honeycomb panels, which are joined together with riveted inner and outer doublers. Linear-shaped charges will be installed on four of the panels, which are hinged at the aft end, to provide a means of exposing the LM and separating the S/M from the SLA.

5-20. MODULE MATING AND FINAL ASSEMBLY.

5-21. Upon completion of structural assembly, the modules are cleaned and sent to a cleanroom for installation and checkout of all systems. The modules are then mated for fit-check and alignment to ensure conformance to design. Alignment is checked optically with theodolites, sight levels, or autocollimators. Each module is also given a weight and balance check to determine its center-of-gravity. Following completion of individual and combined systems checkout, a detailed prelaunch integrated systems check is performed. After assurance that all systems perform according to design criteria, the modules are demated, packaged, and shipped to the designated test site.

APOLLO TRAINING EQUIPMENT

6-1. GENERAL.

6-2. The nature of the Apollo missions demands completely competent personnel for the program. A training program has been established to provide total competence and integration between management, staff, flight crew, flight and ground operations control, and test and operations personnel. Equipment for the training program includes Apollo mission simulators and Apollo systems trainers for various Apollo systems.

6-3. APOLLO MISSION SIMULATORS.

6-4. The Apollo mission simulator (figure 6-1) is a fixed-base training device, capable of simulating the characteristics of space vehicle systems performance and flight dynamics. The simulator provides training of Apollo flight crew members in the operation of spacecraft systems, space navigation, and crew procedures for space missions. In addition to normal spacecraft operation, the AMS simulates malfunctioning systems and degraded systems performance. To extend the simulators to full mission-training capability, telemetry data link, added visual window simulation and waste management have been added.

6-5. Although the Apollo mission simulators are intended to operate independently as full mission trainers for flight crews, they may also be used in an integrated mode with the mission control center (MCC) to simulate the spacecraft and provide flight crew training in conjunction with the operations support personnel operating the MCC and manned space-flight network.

6-6. One mission simulator is installed at MSC, Houston, Texas, and one at the Eastern Test Range, Kennedy Space Center, Florida.

6-7. SYSTEMS TRAINER.

6-8. The Apollo systems trainer complex is comprised of five display trainers, each having its own respective system control console. These training devices are intended to familiarize Apollo project personnel with the functional relationship of spacecraft systems, subsystems and components, the effects of malfunctions, and procedures of system management. The five display trainers are provided for the following spacecraft systems:

- Sequential flow, including the following integrated systems: launch escape, earth landing, emergency detection, and crew safety systems

- Stabilization and control system

- Electrical power system

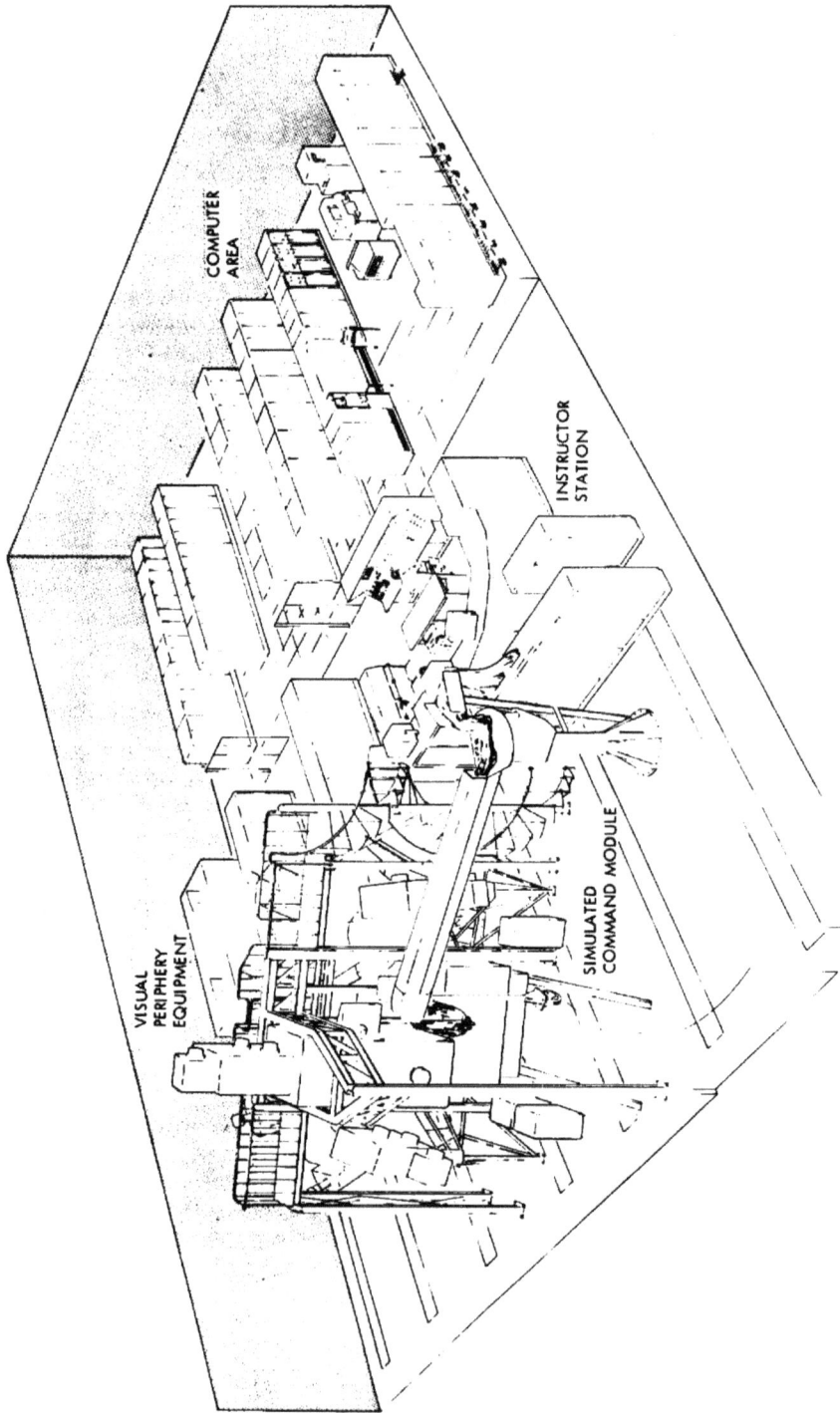

COMPUTER AREA

INSTRUCTOR STATION

SIMULATED COMMAND MODULE

VISUAL PERIPHERY EQUIPMENT

SM-2A-480C

Figure 6-1. Apollo Mission Simulators Installation

- Environmental control system

- Spacecraft propulsion systems (service propulsion and reaction control systems)

6-9. The sequential flow display trainer is capable of accurately displaying a schematic diagram of all sequential operations of the launch escape, earth landing, emergency detection, and crew safety systems. Two panels are incorporated in the trainer to demonstrate normal launch, pad abort, high-altitude abort, early-mission termination, and normal earth landing sequence. Sequence malfunction simulation is limited to circuit interruption, disrupting component operation presented on the panel displays.

6-10. The stabilization and control display trainer presents a functional flow diagram depicting system operation including various switching functions. Simulated spacecraft panels are incorporated in the trainer to simulate the normal operation and malfunction of system components which control and stabilize the spacecraft flight attitude.

6-11. The electrical power system display trainer accurately depicts a schematic flow diagram showing d-c and a-c power distribution to the main buses, including fuel cell and cryogenic storage system flow diagrams. Panels on the trainer demonstrate fuel cell operation, inverter operation, battery recharging, bus switching, and systems management through use of spacecraft panel monitors. Malfunction inputs provide high or low voltage, overload, reverse current, and out-of-tolerance fuel cell monitoring conditions.

6-12. The environmental control system display trainer utilizes two panels to depict, by flow diagram, the pressure suit supply system, oxygen supply system, oxygen cryogenic input system, water-glycol system, and water system during normal ascent operation from the launch pad, while in space, and during entry. Simulated spacecraft panel monitors will be activated to show the pressures and temperatures for the operating modes. A malfunction capability is incorporated in the trainer to indicate high- or low-system parameters and, emergency conditions such as cabin pressure loss and contaminated potable water.

6-13. The propulsion system display trainer utilizes three display panels to present plumbing diagrams of the command module reaction control system, service module reaction control system, and service propulsion system in both manual and automatic modes of operation. Malfunction switches are incorporated in the panels to demonstrate visual malfunction of system components as indicated on spacecraft panel monitors.

APOLLO TEST PROGRAM

7-1. GENERAL.

7-2. This section delineates the test program for the development of Apollo spacecraft. The development program is divided essentially into two blocks, with three interrelated phases: Block I boilerplate and spacecraft missions, Block II spacecraft missions, and propulsion system testing for both blocks. A description of ground support equipment categories and completed Apollo missions is also presented.

7-3. Boilerplates were research and development vehicles which simulated production spacecraft in size, shape, structure, mass, and center of gravity. Each boilerplate was equipped with instrumentation to record mission parameter data for engineering review and evaluation. The data gained from the testing of boilerplate configurations was used in determining production spacecraft flight parameters.

7-4. Spacecraft are production vehicles. These vehicles incorporate numerous modification, flight profile changes, and operating technique revisions deemed necessary as a result of boilerplate mission evaluations. Spacecraft configurations vary in order to meet interface requirements of Saturn V and uprated Saturn I boosters. Variations are made in the spacecraft adapters and the inserts required to satisfy booster interface. C/M and S/M size remain constant.

7-5. Propulsion system testing is accomplished with propulsion system test fixtures. The fixtures are not ground support equipment items, but are unique test platforms for the spacecraft propulsion system. The fixtures are fully instrumented to record engine and propellant system operation through varied operating ranges.

7-6. SPACECRAFT DEVELOPMENT.

7-7. Spacecraft development includes tests and vehicles used for the development of manned spacecraft. The relation between test vechicles, boilerplates, spacecraft, and the Apollo spacecraft development program is shown in figure 7-1.

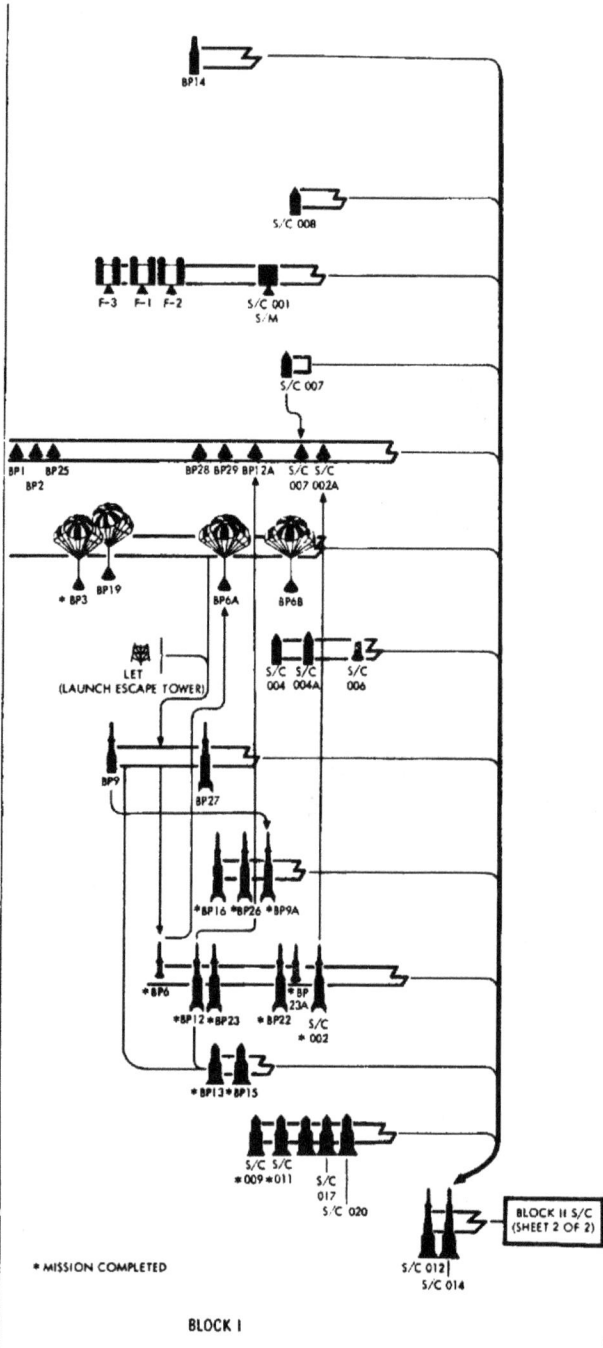

MAJOR GROUND TESTS
(BP14)
HOUSE SPACECRAFT
HARDWARE DEVELOPMENTAL TOOL
VIBRATION AND ACOUSTIC TESTS
(COMPLETED)

ENVIRONMENTAL PROOF TESTS
(THERMAL VACUUM) (S/C 008)
EVALUATE S/C UNDER SIMULATED
MISSION ENVIRONMENTAL CONDITIONS

PROPULSION TESTS (F-1,
F-2, F-3, S/C 001)
SYSTEM COMPATIBILITY TESTS

RECOVERY AND IMPACT TESTS
(BP1, BP2, BP3, BP6A, BP6B, BP12A, BP19,
BP28, BP29, S/C 002A, S/C 007)

MODAL, LAND AND WATER IMPACT TESTS,
AND FLOTATION/UPRIGHTING TESTS

PARACHUTE RECOVERY TESTS

STRUCTURAL TESTS (S/C 004, S/C 004A), S/C 006
VERIFY RIGIDITY AND STRUCTURAL
INTEGRITY UNDER SIMULATED
LOADING CONDITIONS

DYNAMIC TESTS (BP9,
BP27)
DETERMINE STRUCTURAL
COMPATIBILITY WITH LAUNCH
VEHICLES

MICROMETEOROID EXPERIMENT
BP16, BP26, AND BP9A MISSIONS
SUCCESSFUL

ABORT TESTS (BP6, BP12, BP22,
BP23, BP23A, S/C 002,
ABORT CAPABILITIES FOR PAD,
TRANSONIC, HI-ALTITUDE, AND
HI-Q VERIFIED. (BP6, BP12, BP22, BP23,
AND BP23A MISSIONS COMPLETED)

LAUNCH ENVIRONMENT TESTS
(BP13, BP15)
QUALIFY LAUNCH VEHICLES
(BP13 AND BP15 MISSIONS COMPLETED)

UNMANNED FLIGHTS S/C 009,
S/C 011, S/C 017, S/C 020)
SUPERCIRCULAR ENTRY
FLIGHT TO QUALIFY S/C
SYSTEMS & HEATSHIELD
PRIOR TO MANNED FLIGHT

MANNED CONFIGURATION (S/C 012,
S/C 014)
MANNED CONFIGURED FLIGHT TO
DEMONSTRATE OPERATION AND PERFORMANCE
OF S/C AND SYSTEMS.

* MISSION COMPLETED

BLOCK I

SM-2A-576 J

Figure 7-1. Apollo Spacecraft Development Program (Sheet 1 of 2)

BLOCK I S/C
(SHEET 1 OF 2)

RECOVERY TEST VEHICLES
(S/C 2S-1 AND S/C 007A)
WATER AND LAND, IMPACT/FLOTATION
AND POST LANDING TESTS

S/C
2S-1

S/C
007A

STATIC TESTS
S/C 2S-2)
S/M FOR STATIC STRUCTURAL TESTS

S/C
2S-2

ENVIRONMENTAL PROOF VEHICLE
(S/C 2TV-1)
THERMAL VACUUM TESTS

S/C
2TV-1

MANNED CONFIGURATION FLIGHT
(S/C 101 THROUGH 112)
EARTH ORBITAL AND LUNAR
LANDING MISSION

S/C 101
S/C 102
S/C 103
S/C 104
S/C 105
S/C 106
S/C 107
S/C 108
S/C 109
S/C 110
S/C 111
S/C 112

BLOCK II

SM-2A-834B

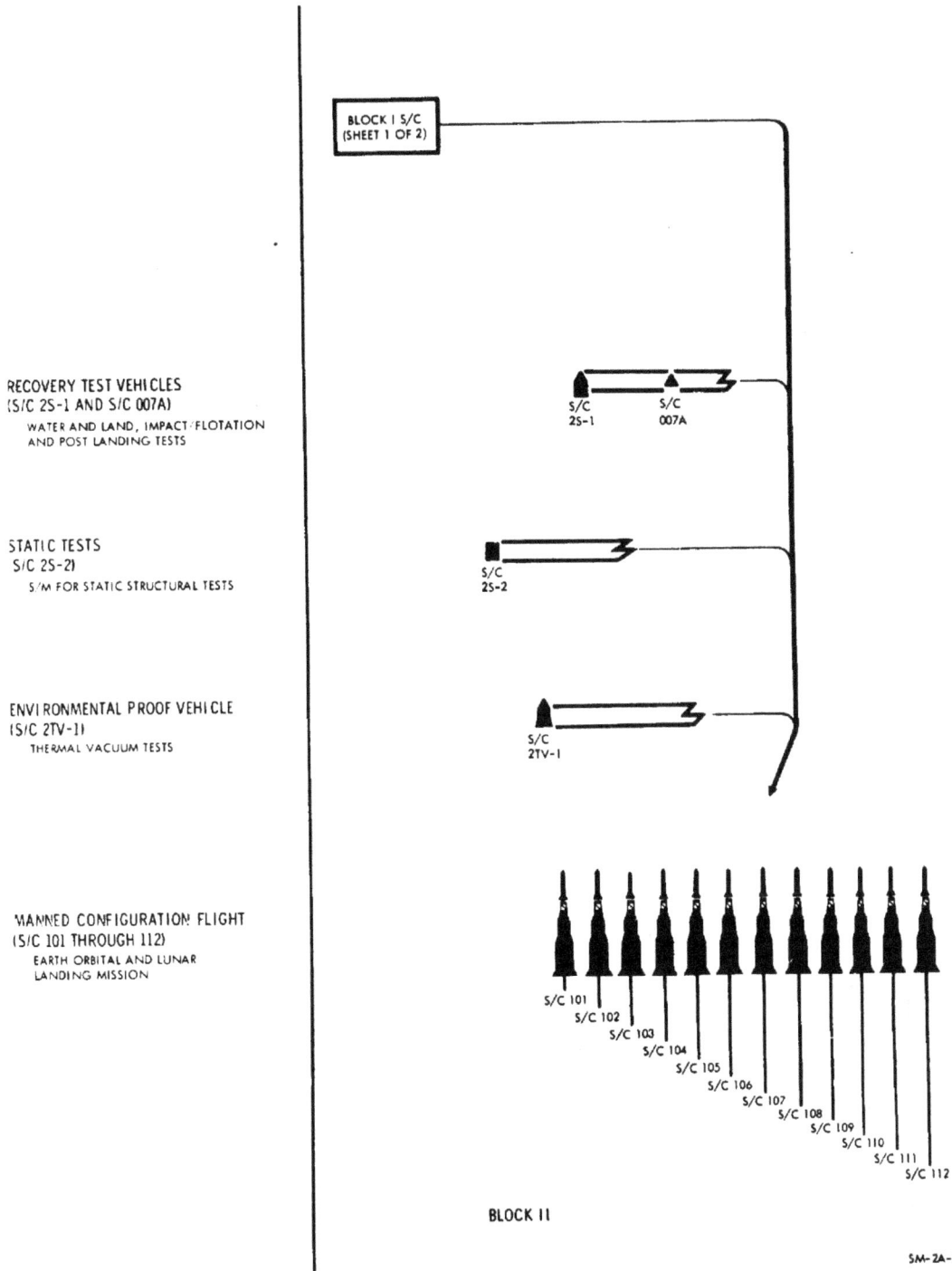

Figure 7-1. Apollo Spacecraft Development Program (Sheet 2 of 2)

SYSTEMS CONFIGURATION LEGEND

- C - COMPLETE
- P - PARTIAL
- R - R & D INSTRUMENTATION EQUIPMENT ONLY
- S - SIMULATED OR INERT
- SP - SPECIAL
- MI - CONTROL PROGRAMER
- MP - APOLLO MISSION PROGRAMER FOR UNMANNED MISSIONS

NOTE: THE INFORMATION IN THIS CHART IS BASED ON THE LATEST PROGRAM PLAN. NO ATTEMPT IS MADE TO SHOW IMPENDING CHANGES.

	BOILERPLATES	TEST SITES	MISSIONS OR PURPOSE	STRUCTURE (Tower, Command Module, Service Module, Boost Protective Cover)	ADAPTER (S-IV, S-IVB, Extension, Special)	LAUNCH VEHICLES (Saturn I, Saturn I, Uprated, Saturn V, Little Joe II)	LAUNCH ESCAPE SYSTEM (LES) (Launch Escape Motor, Pitch Control Motor, Jettison Motor, Abort Sequencer, Canard)	SERVICE PROPULSION SYSTEM (SPS)	SEPARATION SYSTEM (LES - CM, CM - SM, SM - Adapter, Adapter Booster)	EARTH LANDING SYSTEM (ELS) (Parachute, Recovery Aids, Recovery Sequencer, Flotation Equip.)	GUIDANCE & NAVIGATION SYSTEM (G&N)	STABILIZATION & CONTROL SYSTEM (SCS)	REACTION CONTROL SYSTEM (RCS) (Command Module, Service Module)	ENVIRONMENTAL CONTROL SYSTEM (ECS) (CM, SM)	ELECTRICAL POWER SYSTEM (EPS) (CM, SM)	COMMUNICATIONS SYSTEM	DEEP SPACE (S-BAND) EQUIP.	INSTRUMENTATION SYSTEM	CONTROLS AND DISPLAYS	APOLLO MISSION PROGRAMER	CREW COUCHES	SUPPORT & RESTRAINT	CREW SYSTEM	ANTHROPOMORPHIC DUMMIES	LUNAR MODULE
MAJOR GROUND TEST VEHICLES 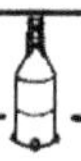	BP-14	DOWNEY, CALIF.	HOUSE SPACECRAFT NO. 1 (HARDWARE DEVELOPMENT TOOL)	C C	C		C	S	P P S S	C C C	C	C	P P	C P	P P	C	P	C	C		C	P	P		
EARTH RECOVERY AND IMPACT TEST VEHICLES 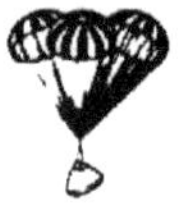	BP-1	DOWNEY, CALIF	COMMAND MODULE FOR LAND AND WATER IMPACT TESTS	C														P			S	S		S	
	BP-2	DOWNEY, CALIF	LAND AND WATER IMPACT TESTS, AND UPRIGHTING SYSTEM DEVELOPMENT TESTS	C						C								P							
	BP-6A (REFURBISHED BP-6)	EL CENTRO, CALIF	PARACHUTE RECOVERY TESTS	C						C C P					R	R		P				P			
	BP-6B (REFURBISHED BP-6A)	EL CENTRO, CALIF	PARACHUTE RECOVERY TESTS	C						C C P					R	R		R							
	BP-3	EL CENTRO, CALIF.	COMMAND MODULE FOR PARACHUTE RECOVERY TESTS (DESTROYED)	C						C C					R	R		P				P			
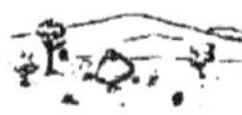	BP-12A (REFURBISHED BP-12)	DOWNEY, CALIF.	WATER IMPACT TEST	C						P															
	BP-19	EL CENTRO, CALIF.	SAME AS BP-3	C						C C P					R	R		P							
	BP-25	MSC	WATER RECOVERY AND HANDLING DEVELOPMENT	C																					
	BP-28	DOWNEY, CALIF.	LANDING IMPACT TEST	C						S	S	S	S	S	S	S	S	S	S & R		S & C	S	S	S	
	BP-29	MSC	FLOTATION TESTS AND QUALIFICATION OF BLOCK I UPRIGHTING SYSTEM	C						P C					P	P									
ABORT TEST VEHICLES 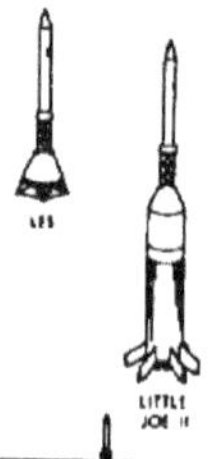 LES, LITTLE JOE II	BP-6 (MISSION COMPLETE)	WSMR	PAD ABORT VEHICLE	C C	C		C C C C		C	C C					P & P	R		P							
	BP-12 (MISSION COMPLETE)	WSMR	VEHICLE FOR TRANSONIC ABORT USING LITTLE JOE II	C C C	C	C	C C C C		C C	C C				SP	R & P	R		P							
	BP-22 (MISSION COMPLETE)	WSMR	VEHICLE FOR HI-ALTITUDE ABORT VERIFICATION OF LES USING LITTLE JOE II	C C C C	C	C	C C C C C		C C	C C				SP	P	R		P							
	BP-23 (MISSION COMPLETE)	WSMR	HIGH-Q ABORT TEST, VERIFICATION OF LES, ELS, CANARD, ETC	C C C C	C	C	C C C C C		C C	C C				SP	P	R		P							
	BP-23A (MISSION COMPLETE)	WSMR	C M FOR PAD ABORT	C C C	C		C C C C C		C C	C C				SP	P	R		P							
LAUNCH ENVIRONMENT VEHICLES 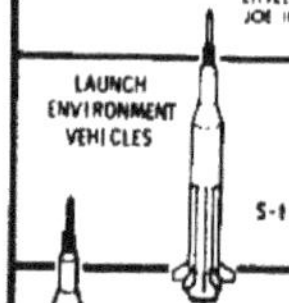 S-I	BP-13 (MISSION COMPLETE)	CAPE KENNEDY	FIRST VEHICLE TO QUALIFY SATURN I	C C C	C C	C	S S C C		C C				S	SP	R & P	R		P							
	BP-15 (MISSION COMPLETE)	CAPE KENNEDY	SECOND VEHICLE TO QUALIFY SATURN I	C C C	C	C	C C C C		C C				S	SP	R & P	R		P							
DYNAMIC TEST VEHICLES 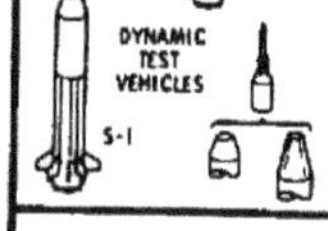 S-I	BP-9	MSFC	DETERMINATION OF DYNAMIC STRUCTURAL COMPATIBILITY WITH SATURN I LAUNCH VEHICLE	C C C	C		S S S S		P P P																
	BP-27	MSFC	DETERMINATION OF DYNAMIC STRUCTURAL COMPATIBILITY WITH UPRATED IB AND V LAUNCH VEHICLES	C C C	C		S S S S		P P P P	S S S	S	S	S S	S S	S S	S	S	S	S		S	S	S		
MICROMETEOROID EXPERIMENT VEHICLES 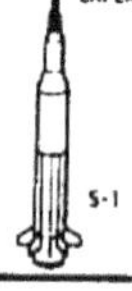 S-I	BP-16 (MISSION COMPLETE)	CAPE KENNEDY	EARTH ORBITAL MICROMETEOROID EXPERIMENT	C C C	C	C C	S S C C		C C				S		P										
	BP-26 (MISSION COMPLETE)	CAPE KENNEDY	SAME AS BP-16	C C C	C	C	S S C C		C C				S		P										
	BP-9A (MISSION COMPLETE)	CAPE KENNEDY	SAME AS BP 16	C C C	C C	C	S S C C		C C				S		P										

SM-2A-577G

Figure 7-2. Block I Boilerplate Vehicle Systems Configuration for Spacecraft Development

SYSTEMS CONFIGURATION LEGEND

- C - COMPLETE
- P - PARTIAL
- P - R & D INSTRUMENTATION EQUIPMENT ONLY
- S - SIMULATED OR INERT
- SP - SPECIAL
- M1 - CONTROL PROGRAMER
- M2 & M3 - APOLLO MISSION PROGRAMER FOR MANNED OR UNMANNED MISSIONS

NOTE: THE INFORMATION IN THIS CHART IS BASED ON THE LATEST PROGRAM PLAN. NO ATTEMPT IS MADE TO SHOW IMPENDING CHANGES.

Vehicles	Spacecraft	Test Sites	Missions or Purpose	Structure (Tower, Command Module, Service Module, Boost Protective Cover)	Spacecraft Adapter (SLA, Extension, Special)	Launch Vehicles (Saturn I, Saturn I/Uprated, Saturn V, Little Joe II)	Launch Escape System (LES) (Launch Escape Motor, Pitch Control Motor, Jettison Motor, Abort-Sequencer, Canard)	Service Propulsion System (SPS)	Separation System (LES - CM, CM - SM, SM - Adapter, LEM - SLA)	Earth Landing System (ELS) (Parachute, Recovery Aids, Recovery Sequencer, Flotation Equip.)	Guidance & Navigation System (G&N)	Stabilization & Control System (SCS)	Reaction Control System (RCS) (Command Module, Service Module)	CM SM Environmental Control System (ECS)	CM SM Electrical Power System (EPS)	Communications System	Deep Space (S-Band) Equip.	Instrumentation System	Controls and Displays	Apollo Mission Programer	Crew Couches	Support & Restraint	Crew System	Anthropomorphic Dummies	Lunar Module (LM)
EARTH RECOVERY AND IMPACT TEST VEHICLES 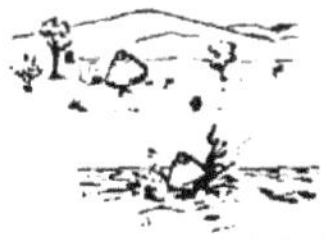	S/C 002A	DOWNEY, CALIF.	S/C 002 REFURBISHED FOR LAND IMPACT TESTS	C																	C	C		C	
	S/C 007	DOWNEY, CALIF. AND MSC	SPACECRAFT FOR ACOUSTIC, WATER IMPACT AND POST LANDING TESTS	SP C C S				S	S S S	S S S C P C C	S	S	S S	S S P	S S P	S P	S	P P	S P		S C	S C	S P	S C	
ABORT TEST VEHICLES 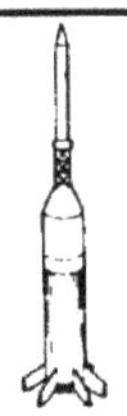	S/C 002	WSMR	POWER ON TUMBLING ABORT	C C C C	C	C	C C C C C		C C	C C P			S		P	P		P							
STRUCTURAL TEST VEHICLES 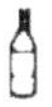	S/C 004	DOWNEY, CALIF.	STATIC STRUCTURAL TESTS	SP C C SP	C			S	P P P	P															
	S/C 004A	DOWNEY, CALIF.	STATIC AND THERMAL STRUCTURAL TESTS	C					P																
	S/C 006	DOWNEY, CALIF.	C/M FOR ELS LOAD TESTS AND LET SEPARATION TESTS	C C					P																
ENVIRONMENTAL PROOF VEHICLE 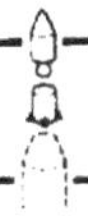	S/C 008	MSC	EVALUATE S/C UNDER SIMULATED MISSION ENVIRONMENTAL CONDITIONS THERMAL VACUUM TESTS	C C	C			S	S S	C P C P	C	C	P C	C C	C C	C	P	C & SP	C		C	C	C		
PROPULSION TEST VEHICLES	S/C 001	(PSDF) WSMR	COMPATIBILITY VERIFICATION OF S/C SERVICE MODULE PROPULSION SYSTEMS WHEN INTEGRATED WITH S/M SUBSYSTEM	C	C			C	P			SP	C	P	C			P							
UNMANNED FLIGHT VEHICLES 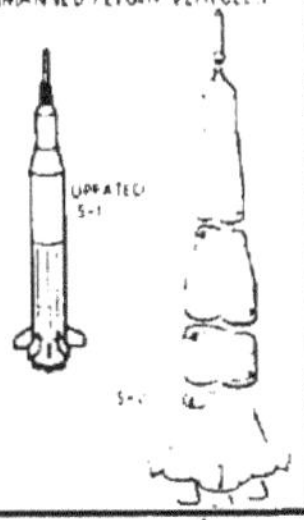	S/C 009	CAPE KENNEDY	SUPER CIRCULAR ENTRY FLIGHT TO QUALIFY S/C PRIOR TO MANNED FLIGHT	C C C C	C	C	C C C C	C	C C C	C C C C		P	C C	P P	P P	P		P	P	M1					
	S/C 011	CAPE KENNEDY	SUPER CIRCULAR ENTRY FLIGHT TO DEMONSTRATE OPERATION AND PERFORMANCE OF S/C AND SYSTEMS	C C C C	C	C	C C C C	C	C C C	C C C C	C	P	C C	C C	C C	C	P	C	C	M3					
	S/C 017	CAPE KENNEDY	SIMULATED LUNAR RETURN ENTRY	C C C C	C C	C	C C C C	C	C C C C	C C C C	P	P	C C	C C	C C	P	P	P	P	M3					P
	S/C 020	CAPE KENNEDY	SIMULATED LUNAR RETURN ENTRY	C C C C	C C	C	C C C C	C	C C C C	C C C C	P	P	C C	C C	C C	P	P	P	P	M3					P
MANNED CONFIGURATION VEHICLES FLIGHT 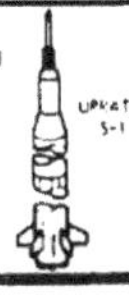	S/C 012	CAPE KENNEDY	EARTH ORBITAL FLIGHT & SEA RECOVERY TO EVALUATE IN-FLIGHT OPERATIONS	C C C C	C	C	C C C C	C	C C C	C C C C	C	C	C C	C C	C C	C	P	C	C	M2	C	C	C		
	S/C 014	CAPE KENNEDY	EARTH ORBITAL FLIGHT TO EVALUATE IN-FLIGHT CSM OPERATIONS	C C C C	C	C	C C C C	C	C C C	C C C C	C	C	C C	C C	C C	C	P	C	C	M2	C	C	C		

SM-2A-378G

Figure 7-3. Block I Spacecraft Vehicle Systems Configuration for Spacecraft Development

7-8. Boilerplate and spacecraft vehicle systems configuration for spacecraft development of Block I and Block II vehicles is shown in figures 7-2 through 7-4. Letters are used to designate the following: a complete system (C), a partial system (P), R&D instrumentation equipment only (R), a simulated or inert system (S), a special system (SP), and different configurations of the Apollo mission programer (M1, M2, or M3). A blank space in any column indicates the described system is not installed.

7-9. BLOCKS I AND II.

7-10. A block concept is used for spacecraft development to separate the vehicles into different phases, such as research and development (Block I) and production vehicles for earth orbital and lunar missions (Block II). Paragraphs 7-11 through 7-14 give a breakdown of Blocks I and II and their functions.

NOTE

Block II information is based on preliminary data only.

7-11. Block I encompasses the entire boilerplate program, and spacecraft 001, 002, 002A, 004, 004A, 006, 007, 008, 009, 011, 012, 014, 017, and 020.

7-12. The boilerplate portion of Block I provide:

a. Early support of systems development for land impact, water impact, and parachute recovery prequalification tests

b. Systems qualification to support spacecraft programs including pad abort, high-altitude abort, and house spacecraft No. 1 (boilerplate 14) which contains all systems

c. Marshall Space Flight Center support including Saturn I development and micro-meteoroid detection

d. Space-flight capabilities development and coordination of manufacturing, testing operations, engineering, and NASA functions.

7-13. The spacecraft portion of Block I provides:

a. Command module and service module development for manned earth orbital missions

b. Demonstration of operational capabilities of systems including all types of aborts, land recovery, water recovery, uprated Saturn I and Saturn V operation (and compatibility), and operation during earth orbits (unmanned)

c. Qualified teams development for checkout, launch, manned space flight network, recovery, and flight analysis.

NOTE

S/C 007 will be refurbished and designated 007A for
Block II postlanding tests.

7-14. Block II encompasses spacecraft 2S-1, 007A, 2S-2, 2TV-1, 101, 102, 103, 104, 105, 106, 107, 108, 109, 110, 111, and 112, and provides:

a. Incorporation of lunar module provisions

b. Improvement of center-of-gravity in command module

c. Evaluation and incorporation of system changes with respect to lunar mission and reliability impact.

SYSTEMS CONFIGURATION LEGEND: C - COMPLETE; P - PARTIAL; S - SIMULATED OR INERT; SP - SPECIAL	NOTE: THE INFORMATION IN THIS CHART IS BASED ON PRELIMINARY DATA. NO ATTEMPT IS MADE TO SHOW IMPENDING CHANGES. SPACECRAFT NUMBER	TEST SITES	MISSIONS OR PURPOSE	STRUCTURE (Tower & Boost Protective Cover, Command Module, Service Module)	SPACECRAFT ADAPTER (SLA, Special)	LAUNCH VEHICLE (Saturn UPRATED, Saturn V)	LAUNCH ESCAPE SYSTEM (Launch Escape Motor, Pitch Control Motor, Jettison Motor, Canard)	SERVICE PROPULSION SYSTEM (SPS)	SEPARATION SYSTEM (LES - CM, CM - SM, SM - Adapter, LEM - SLA)	SEQUENCE CONTROLLERS (Master Event, Earth Landing, SM-Separation, CM-RCS, Lunar Docking Events, LEM-SLA Separation)	RECOVERY SYSTEMS (ELS) (Earth Landing, Recovery Aids, Flotation Equip.)	GUIDANCE & NAVIGATION SYSTEM (G&N)	STABILIZATION & CONTROL SYSTEM (SCS)	REACTION CONTROL SYSTEM (RCS) (Command Module, Service Module)	ENVIRONMENTAL CONTROL SYSTEM (ECS) (CM, SM)	ELECTRICAL POWER SYSTEM (EPS) (CM, SM)	COMMUNICATIONS SYSTEM (Deep Space, Near Earth, Recovery)	INSTRUMENTATION SYSTEM (Operational, R & D, Scientific)	CONTROLS AND DISPLAYS		EMERGENCY DETECTION SYSTEM	CREW EQUIPMENT (Couches And Restraints, Accessories & Provisions, Anthropomorphic Dummies)	LUNAR MODULE	LEM DOCKING SYSTEM
MAJOR GROUND TEST VEHICLES 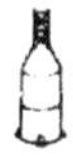																								
RECOVERY TEST VEHICLES	2S-1	DOWNEY, CALIF.	MODAL, ACOUSTIC, FLOTATION AND IMPACT	C C	C			C	S S S	S S S S S S	S S C	S	S	S C	S S	S S	S S	P	S			C S C	S	P
	007A	DOWNEY, CALIF.	S/C 007 REFURBISHED FOR BLOCK II POST LANDING TESTS	C						S S S S S	P C C	S	S	S	P	P	S C	P	P			C P C		
STATIC TESTS	2S-2	DOWNEY, CALIF.	S/M FOR STATIC TESTS	SP & P C C	C				P P P									P	S					P
ENVIRONMENTAL PROOF VEHICLE	2TV-1	MSC	THERMAL VACUUM TESTS	C C	C			S	S S S S	C C C C C C	C C C	C	C	S S	C C	C C	S S	C P P	C		C	C C C	C	C
MANNED FLIGHTS 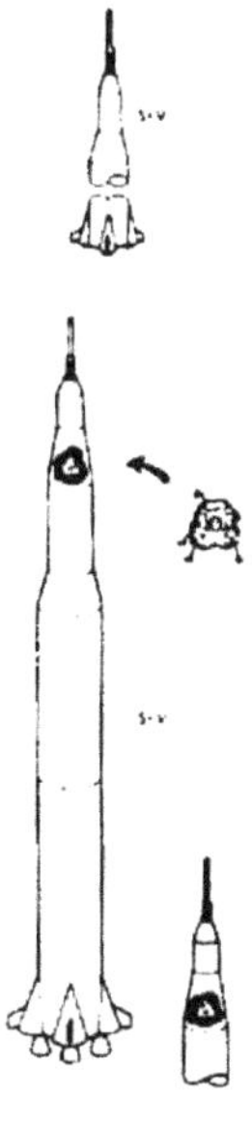	101	CAPE KENNEDY	ELLIPTICAL EARTH ORBITAL FLIGHT TO EVALUATE RENDEZVOUS AND DOCKING MANEUVERS	C C C	C	C	C C C C	C	C C C C	C C C C C C	C C C	C	C	C C	C C	C C	C C C	C P P	C		C	C C	C	C
	102	CAPE KENNEDY	SMALL ELLIPTICAL EARTH ORBITAL FLIGHT TO EVALUATE AND CSM RENDEZVOUS AND DOCKING	C C C	C	C	C C C C	C	C C C C	C C C C C C	C C C	C	C	C C	C C	C C	C C C	C P	C		C	C C	C	C
	103	CAPE KENNEDY	LUNAR LANDING RESEARCH AND DEVELOPMENT	C C C	C	C	C C C C	C	C C C C	C C C C C C	C C C	C	C	C C	C C	C C	C C C	C P	C		C	C C	C	C
	104	CAPE KENNEDY	LUNAR LANDING TO EVALUATE MAN ON LUNAR SURFACE AND DEMONSTRATE CAPABILITY	C C C	C	C	C C C C	C	C C C C	C C C C C C	C C C	C	C	C C	C C	C C	C C	C P	C		C	C C	C	C
	105	CAPE KENNEDY	LUNAR LANDING	C C C	C	C	C C C C	C	C C C C	C C C C C C	C C C	C	C	C C	C C	C C	C C	C P	C		C	C C	C	C
	106	CAPE KENNEDY	LUNAR LANDING	C C C	C	C	C C C C	C	C C C C	C C C C C C	C C C	C	C	C C	C C	C C	C C	C P	C		C	C C	C	C
	107	CAPE KENNEDY	LUNAR LANDING	C C C	C	C	C C C C	C	C C C C	C C C C C C	C C C	C	C	C C	C C	C C	C C	C P	C		C	C C	C	C
	108	CAPE KENNEDY	LUNAR LANDING	C C C	C	C	C C C C	C	C C C C	C C C C C C	C C C	C	C	C C	C C	C C	C C	C P	C		C	C C	C	C
	109	CAPE KENNEDY	LUNAR LANDING	C C C	C	C	C C C C	C	C C C C	C C C C C C	C C C	C	C	C C	C C	C C	C C	C P	C		C	C C	C	C
	110	CAPE KENNEDY	LUNAR LANDING	C C C	C	C	C C C C	C	C C C C	C C C C C C	C C C	C	C	C C	C C	C C	C C	C P	C		C	C C	C	C
	111	CAPE KENNEDY	LUNAR LANDING	C C C	C	C	C C C C	C	C C C C	C C C C C C	C C C	C	C	C C	C C	C C	C C	C P	C		C	C C	C	C
	112	CAPE KENNEDY	LUNAR LANDING	C C C	C	C	C C C C	C	C C C C	C C C C C C	C C C	C	C	C C	C C	C C	C C	C P	C		C	C C	C	C

SM-2A-579J

Figure 7-4. Block II Spacecraft Vehicle Systems Configuration for Spacecraft Development

7-15. The primary differences between the systems of Block I and Block II manned spacecraft are listed as follows:

Spacecraft System	Block I	Block II
ELS	Nylon main parachute risers.	Steel cable main parachute risers.
ECS		Redundant coolant loop added.
		LM pressurization controls added to C/M.
EPS	S/M separation batteries A and B, used to separate C/M-S/M.	No S/M separation batteries — capabilities for C/M-S/M separation added to fuel cells.
		One flight bus (d-c) added.
		No frequency meter on MDC.
RCS	No fuel dump capability in C/M.	Rapid fuel dump C/M-RCS.
G&N	Interfaces with SCS.	Completely independent of SCS.
	Long-relief-eyepieces installed by astronauts.	Long-relief-eyepieces incorporated into sextant and telescope.
	CDUs electro-mechanical.	CDUs all electronic.
		AGC size reduced and memory capacity increased.
		Reduced size and weight of IMU.
		Redesigned navigation base.
SCS	Interfaces with G&N.	
	One FDAI	Two FDAIs
	AGCU	Gyro display coupler (GDC) in place of AGCU.
	Rate gyro assembly	BMAG assembly in place of RGA.
Crew system	No portable life support system.	Portable life support system (PLSS).
	Three one-man liferafts.	One three-man liferaft.
	Fluorescent illumination.	Electroluminescent, flourescent, and incandescent illumination.
	No EVT capabilities.	Extravehicular transfer (EVT) and tethering capabilities.

Spacecraft System	Block I	Block II
T/C	C-band transponder.	C-band transponder in S/C 101 and 102
	One S-band transponder.	Two S-band transponders.
	One S-band power amplifier.	Two S-band power amplifiers.
	No high-gain antenna.	High-gain antenna and controls.
		Additional VHF/AM capabilities.
	No rendezvous transponder and antenna.	Rendezvous radar transponder and antenna.
Docking system	None	Installed in Block II S/C only.

7-16. BOILERPLATE MISSIONS.

7-17. The boilerplate missions were primarily research and development tests to evaluate the structural integrity of the spacecraft and confirm basic engineering concepts relative to system performance and compatibility. A number of missions were conducted during this phase of the test program. The missions were scheduled to follow a pattern of development starting with basic structure evaluation, followed by systems performance and compatibility confirmation.

7-18. Each mission was dependent upon the previous mission in developing the systems and operations requisite for lunar exploration. Prior to the start of any mission, the boilerplate to be tested was thoroughly checked at the manufacturer test preparation area under the direction of NASA inspectors. After system and structural checkout was approved, the boilerplate was shipped to the test site for further checkout and mating. A launch countdown was started only after the second checkout and mating had been approved. Figure 7-5 depicts a water impact test.

7-19. BLOCK I BOILERPLATE TEST PROGRAM. The following is a list of each boilerplate and relative mission data. Boilerplates and their missions are part of the Block I portion of the Apollo program. The list is intended as a cross-reference for boilerplate objectives. Chronological order of missions and test grouping is not intended or reflected in the arrangement of the list.

Boiler-plate No.	Test Site	Purpose	Mission	Launch Vehicle
BP1	Downey, Calif.	Development and evaluation of crew shock absorption system; evaluation of C/M on land and water, during and after impact.	Drop tests utilizing impact facility at Downey, Calif.	None
BP2	Downey, Calif.	Same as BP1 and development of the uprighting system.	Drop tests and uprighting tests utilizing impact facility at Downey, Calif.	None

Block I Boilerplate Test Program (Cont)

Boiler-plate No.	Test Site	Purpose	Mission	Launch Vehicle
BP3	El Centro, Calif.	To evaluate parachute recovery system in the air (destroyed).	Parachute recovery via air drop.	Aircraft (drop)
BP6	White Sands Missile Range (WSMR), New Mexico	To determine aerodynamic stability, tower vibration, and spacecraft dynamics during a pad abort; to demonstrate capability of LES to propel C/M to a safe distance from launch area during a pad abort.	Pad abort mission successfully completed 7 November 1963.	Launch escape system motor
BP6A	El Centro, Calif.	BP6 refurbished for parachute recovery system test in the air.	Parachute recovery system evaluation via air drop.	Aircraft (drop)
BP6B	El Centro, Calif.	BP6A refurbished for parachute recovery system tests.	Parachute recovery system evaluation tests.	Aircraft (drop)
BP9	Marshall Space Flight Center (MSFC), Alabama	Dynamic test to be determined by National Aeronautics and Space Administration.	Determination of dynamic structural compatibility of test S/C with Saturn I.	None
BP9A	Kennedy Space Center, Florida	Utilized for launch vehicle qualification and micrometeoroid experiment.	Micrometeoroid experiment. Successfully launched into orbit 30 July 1965.	Saturn I
BP12	WSMR, New Mexico	To determine aerodynamic stability characteristics of Apollo escape configuration during transonic abort from Little Joe II. To demonstrate capability of LES to propel C/M to safe distance from launch vehicle during an abort at high-dynamic pressure. To demonstrate structural integrity of launch escape tower and operational characteristics during a transonic abort, and demonstrate spacecraft-Little Joe II compatibility.	Transonic abort mission successfully completed 13 May 1964.	Little Joe II

Block I Boilerplate Test Program (Cont)

Boiler-plate No.	Test Site	Purpose	Mission	Launch Vehicle
BP12A	Downey, Calif.	BP12 refurbished to evaluate C/M during hard rollover water landing condition.	Water impact test utilizing impact facility at Downey, Calif.	None
BP13	Kennedy Space Center, Florida	To qualify Saturn I launch vehicle, demonstrate physical compatibility of launch vehicle and spacecraft under preflight and flight conditions, and determine launch and exit environmental parameters to verify design criteria.	Launch environment mission successfully completed 28 May 1964 to 31 May 1964.	Saturn I
BP14	Downey, Calif.	Developmental tool (house spacecraft No. 1) for use in developing spacecraft systems and preliminary checks in integrated systems compatibility.	Research and developmental tool for systems evaluation (static vehicle).	None
BP15	Kennedy Space Center, Florida	Second boilerplate flown for environmental data. The purpose of this vehicle and BP13 are similar.	Launch exit environment (orbital flight trajectory) mission successfully completed 18 September 1964 to 22 September 1964.	Saturn I
BP16	Kennedy Space Center, Florida	To be utilized for launch vehicle qualification and micrometeoroid experiment.	Micrometeoroid experiment. Successfully launched into orbit 16 February 1965.	Saturn I
BP19	El Centro, Calif.	To evaluate parachute recovery system in the air.	Parachute recovery system evaluation through means of command module air drop.	Aircraft (drop)
BP22	WSMR, New Mexico	To verify LES, ELS, and canard system during high-altitude abort.	Qualify LES, ELS sequence timing during abort. Mission completed 19 May 1965.	Little Joe II

Block I Boilerplate Test Program (Cont)

Boiler-plate No.	Test Site	Purpose	Mission	Launch Vehicle
BP23	WSMR, New Mexico	Verification of LES and ELS during high-Q abort.	Demonstration of launch escape vehicle structural integrity and recovery of C/M following high-Q abort. Successfully completed mission 8 December 1964.	Little Joe II
BP23A	WSMR, New Mexico	To qualify LES, ELS sequence timing, canard system, dual drogue parachutes, and boost protective cover during pad abort test. BP23 C/M refurbished for pad abort test.	C/M for pad abort evaluation. Mission completed 29 June 1965.	Launch escape system motor.
BP25	Houston, Texas	To demonstrate pickup and handling techniques as required by National Aeronautics and Space Administration.	Demonstration of equipment and handling capability for command module at a site (simulated) recovery as a design for pickup equipment.	None
BP26	Kennedy Space Center, Florida	To be utilized for launch vehicle qualification and micrometeoroid experiment.	Micrometeoroid experiment using NASA-installed equipment. Successfully launched into orbit 25 May 1965.	Saturn I
BP27	MSFC, Alabama	Second dynamic test. Objectives for this ground test will be determined by National Aeronautics and Space Administration.	Determination of dynamic structural compatibility of test S/C with uprated Saturn I and Saturn V launch vehicles.	Uprated Saturn I and Saturn V (Captive test firing)

Block I Boilerplate Test Program (Cont)

Boiler-plate No.	Test Site	Purpose	Mission	Launch Vehicle
BP28	Downey, Calif.	Test vehicle will be impacted (land and water) a number of times in order to evaluate loads imposed on structure due to landing impact and acceleration, accelerations and onset rates imposed on crew by the crew couch attenuation system, stability, and dynamics of the vehicle.	Definition of landing problems by determination and evaluation of loads imposed on structure due to landing impact, and acceleration and onset rates imposed on crew by crew couch attenuation system.	None
BP29	MSC Houston, Texas	To determine flotation characteristics of command module and to qualify Block I uprighting system.	Full-scale flotation and recovery tests for simulated entry and abort conditions.	None

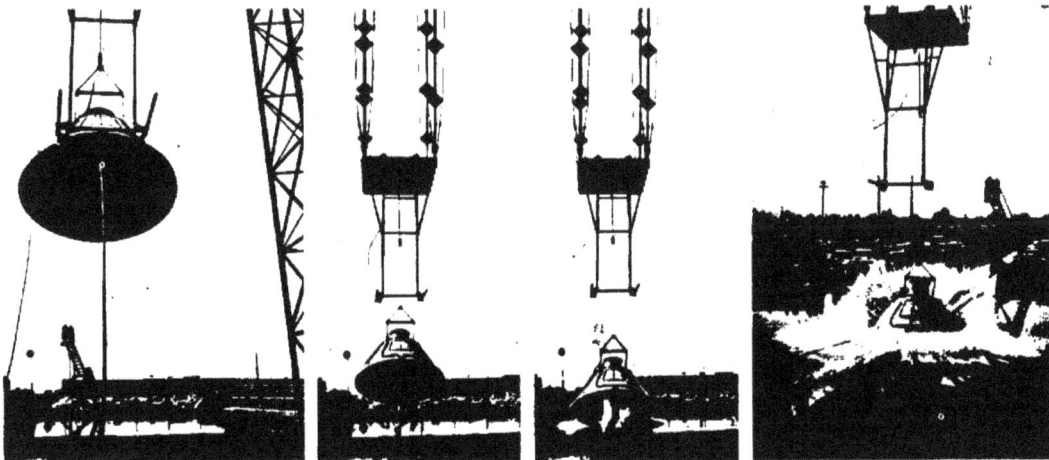

A PENDULUM FOR APOLLO—An impact test facility, from which NASA's unmanned Apollo test command modules are swung and dropped on land or water, is seen in operation at Downey, Calif. The Apollo command module is suspended below the steel platform and the huge "arm" swings the capsule, releasing it at controlled angles and speeds to simulate impact which later manned Apollo spacecraft will undergo upon return to earth.

SM-2A-488A

Figure 7-5. Structural Reliability Test

7-20. SPACECRAFT MISSIONS.

7-21. Spacecraft missions, conducted with production spacecraft, are of a multipurpose nature. The initial missions were conducted to verify production spacecraft structural integrity, and systems operation and compatibility. After the spacecraft structure and systems test missions were completed, a series of unmanned missions were conducted to confirm spacecraft-launch vehicle compatibility and evaluate prelaunch, launch, mission, and post-mission operations. Manned spacecraft missions will be conducted to improve performance and confirm capability between man and spacecraft.

7-22. Each spacecraft mission is dependent upon the success of the previous mission, as an overall program of progressive, manned, space penetration is planned to afford the flight crew maximum familiarity with production spacecraft maneuvers and docking techniques with the LM. Manned missions will penetrate deeper into space as the program progresses. Manned space-flight network techniques will be employed during all earth orbital missions. The system compatibility with Apollo spacecraft will be determined during this phase of the test program. The knowledge gained by the flight crew and engineering personnel during this phase will be analyzed for use during manned lunar surface exploration.

7-23. BLOCK I SPACECRAFT TEST PROGRAM. The following is a complete list of Apollo spacecraft, their missions, and relative data required to complete the Block I portion of the Apollo program. The list is similar in intent to that of paragraph 7-19; the chronological order of spacecraft missions is not intended or reflected in the arrangement of the list.

Space-craft No.	Test Site	Purpose	Mission	Launch Vehicle
001	Propulsion System Development Facility (PSDF), White Sands, Missile Range (WSMR), New Mexico	To verify compatibility of space-craft S/M propulsion system; evaluate S/M service propulsion system and reaction control system during malfunction, normal, and mission profile conditions. To evaluate interface compatibility between all onboard systems during integrated systems test; evaluate performance and compatibility pertaining to ground support equipment, safety rules, operating techniques, and checkout of applicable systems.	Determine spacecraft structural vibration and acoustic characteristics during all phases of SPS operation, and performance of electrical power system and cryogenic storage subsystem during service propulsion and reaction control systems operation.	None (See figure 7-6.)
002	WSMR, New Mexico	To demonstrate structural integrity of production C/M under high dynamic pressure at transonic speed. To determine operational characteristics during power on tumbling abort.	Abort at high dynamic pressures in transonic speed range. Mission successfully completed 20 Jan 1966	Little Joe II

Block I Spacecraft Test Program (Cont)

Space-craft No.	Test Site	Purpose	Mission	Launch Vehicle
002A	Downey, Calif.	S/C 002 refurbished for land impact tests to verify structural integrity of C/M and assure acceptable crew accelerations during land impact.	Land impact tests utilizing impact facility at Downey, Calif.	None
004	Downey, Calif.	S/C for verification of structural integrity and intramodular compatibility of combined module structures under critical loadings.	Static tests.	None
004A	Downey, Calif.	Verification of structural integrity and intramodular compatibility of combined module structures under critical loading.	Static and thermal structural tests.	None
006	Downey, Calif.	C/M ELS load tests and C/M-LET separation tests.	Systems evaluation.	None
007	Downey, Calif. and MSC	This spacecraft will serve a dual purpose: module transmissibility and water impact and flotation tests. Module transmissibility test will determine free-fall lateral bending modes, and longitudinal and shell modes, utilizing two spacecraft configurations. Configuration A will incorporate launch escape tower and C/M; configuration B will incorporate C/M and S/M. Water impact and flotation tests will utilize C/M only. Purpose of water impact test is to verify structural integrity of C/M and crew support system dynamics under shock conditions at water impact. Flotation tests will demonstrate C/M flotation and water-tight integrity in varying sea conditions as well as crew survival in a closed module, and crew egress in varying sea conditions.	Acoustic, water impact, and post-landing tests.	None
008	MSC, Houston, Texas and Downey, Calif.	Spacecraft 008 will undergo both manned and unmanned deep-space environmental control tests. The first of these tests will be conducted at Downey, Calif. After	Thermal vacuum tests.	None

Block I Spacecraft Test Program (Cont)

Space-craft No.	Test Site	Purpose	Mission	Launch Vehicle
		operational checkout of installed systems, spacecraft will be shipped to MSC facility at Houston, Texas, for evaluation and verification of complete spacecraft design under launch simulation, orbital mission simulation, thermal investigation, system failure increments, emergency operations, module separation, entry separation, and recovery aids checkout.		
009	Kennedy Space Center, Florida	To evaluate heat shield ablator performance, RCS and SPS operations, SCS operation, open loop EDS and partial EPS operation. To determine loading separation characteristics and communications performance. To demonstrate operation of recovery system, launch vehicle and spacecraft compatibility. A mission programer, model M1, controlled S/C in-flight operations.	Supercircular high-heat rate entry flight.

Mission successfully completed 26 Feb 1966 | Uprated Saturn I |
| 011 | Kennedy Space Center, Florida | An unmanned flight to evaluate heat shield ablator performance, EDS performance, and SPS propellant retention device. Determine SLA structural loading, separation characteristics, performance of G&N, SCS, ECS, EPS, RCS, and telecommunications. To demonstrate launch vehicle and S/C compatibility, structural integrity, C/M entry, multiple SPS restarts, and recovery. S/C was controlled by model M3 mission programer. | A supercircular high-heat load entry flight.

Mission successfully completed 25 Aug 1966 | Uprated Saturn I |

Block I Spacecraft Test Program (Cont)

Space-craft No.	Test Site	Purpose	Mission	Launch Vehicle
012	Kennedy Space Center, Florida	A manned configured flight to evaluate crew-S/C compatibility, crew tasks and subsystems performance, manual, and backup modes of subsystem separations. To demonstrate closed loop EDS.	An open-end orbital flight for CSM subsystem performance.	Uprated Saturn I
014	Kennedy Space Center, Florida	A manned configured flight to evaluate in-flight CSM performance. To determine radiation levels and to demonstrate closed loop lifting entry.	Elliptical orbital flight for CSM operation.	Uprated Saturn I
017	Kennedy Space Center, Florida	An unmanned flight to evaluate mission programer performance, ECS entry performance, heat shield performance, countdown operations, Saturn V performance, MSFN ability, SCS entry performance, G&C during entry, and G&C boost monitor. To determine open-loop EDS performance, boost environment, structural loading, SPS performance, radiation levels ECS, and EPS operation. To demonstrate structural performance, G&C effectiveness, sea recovery compatibility, and parachute recovery.	Structural integrity and simulated lunar return high-heat rate entry.	Saturn V
020	Kennedy Space Center, Florida	An unmanned flight to evaluate heat shield performance, countdown, launch vehicle repeatability, MSFN ability, LM propulsion performance, and LM G&C stability. To determine open-loop EDS performance, ECS entry performance, LM G&C control, and LM subsystem performance. To demonstrate LM performance, SPS performance, LM fluid systems, LM separation, and sea recovery.	Simulated lunar return high-heat load entry and LM propulsion.	Saturn V

7-24. BLOCK II SPACECRAFT TEST PROGRAM. The following is a complete list of Apollo spacecraft, their missions, and relative preliminary data required for the Block II portion of the Apollo program.

Space-craft No.	Test Site	Purpose	Mission	Launch Vehicle
2S-1	Downey, Calif.	Water and land impact tests to verify structural integrity of C/M and assure acceptable crew accelerations during land impact.	Water and land impact tests utilizing impact facility at Downey, Calif.	None
007A	Downey, Calif.	S/C 007 refurbished for Block II postlanding tests.	Recovery test vehicle.	None
2S-2	Downey, Calif.	To verify structural integrity of CSM	CSM for static structural tests.	None
2TV-1	MSC, Houston, Texas	To evaluate S/C under simulated mission environmental conditions.	Environmental proof (thermal vacuum) tests.	None
101	Kennedy Space Center, Florida	To evaluate rendezvous maneuvers, rendezvous radar transponder, CSM guidance and control, and RCS plume effects. To determine LM propulsion effects. To demonstrate G&C entry, LM ECS operation, transposition and docking, LM ΔV to CSM, landing gear deployment, crew transfer, and one-man LM operation.	Systems evaluation open-end elliptical manned earth orbital flight. Dual launch S/C 101 AS207, LM2 AS 208	Uprated Saturn I
102	Kennedy Space Center, Florida	To evaluate SCS, LM, and CSM in deep space. To determine LM restart effects, SPS effectiveness, plume effects, and heat shield performance. To demonstrate one man operation of CSM and manual entry.	Manned, small elliptical, open-end earth orbital flight.	Saturn V

Block II Spacecraft Test Program (Cont)

Space-craft No.	Test Site	Purpose	Mission	Launch Vehicle
103	Kennedy Space Center, Florida	Research and development to evaluate LM and CSM operations, man on lunar surface, and to demonstrate LM capability.	Manned lunar landing flight.	Saturn V
104	Kennedy Space Center, Florida	Lunar landing.	Manned lunar landing flight.	Saturn V
105	Kennedy Space Center, Florida	Lunar landing.	Manned lunar landing flight.	Saturn V
106	Kennedy Space Center, Florida	Lunar landing.	Manned lunar landing flight.	Saturn V
107	Kennedy Space Center, Florida	Lunar landing.	Manned lunar landing flight.	Saturn V
108	Kennedy Space Center, Florida	Lunar landing.	Manned lunar landing flight.	Saturn V
109	Kennedy Space Center, Florida	Lunar landing.	Manned lunar landing flight.	Saturn V
110	Kennedy Space Center, Florida	Lunar landing.	Manned lunar landing flight.	Saturn V
111	Kennedy Space Center, Florida	Lunar landing.	Manned lunar landing flight.	Saturn V
112	Kennedy Space Center, Florida	Lunar landing.	Manned lunar landing flight.	Saturn V

7-25. TEST FIXTURES.

7-26. Three service propulsion engine test fixtures will be used to test the service module service propulsion systems. A test fixture (figure 7-6) is a structure used for system predevelopment and developmental tests leading to design of a spacecraft article. The fixtures are designated F-1, F-2, and F-3.

7-27. The F-1 test fixture functions to provide a test bed for the service propulsion engine vendor-acceptance, reliability, and qualification tests. The fixture will be used to evaluate the engine for safe operation and performance, to evaluate service propulsion engine basic design parameters, perform early evaluation of propellant system interaction effects, and to evaluate overall compatibility of the propulsion system components and subsystems.

7-28. The F-2 test fixture is a boilerplate structure that simulates the service module service propulsion system. This fixture will be used at PSDF, WSMR, New Mexico, to evaluate the service propulsion system under normal-design limit, and mission flight conditions through hot-propulsion static ground tests. The fixture will be used to permit continuance of the service propulsion system test program during periods when the propulsion spacecraft is out of service for modification, maintenance, and malfunction simulation.

7-29. The F-3 test fixture will be used at AEDC for vendor-acceptance and reliability tests. The F-3 fixture will also be used for engineering system development and static checkout tests by Space and Information Systems Division of North American Aviation, Inc.

7-30. GROUND SUPPORT EQUIPMENT.

7-31. Ground support equipment (GSE), required for Apollo spacecraft, is separated into four categories: checkout, auxiliary, servicing, and handling. The purpose of GSE is to provide the Apollo program with a GSE system that will establish a level of confidence in the onboard spacecraft systems and will ensure mission success within prescribed reliability factors.

7-32. CHECKOUT EQUIPMENT. Checkout equipment consists of acceptance checkout equipment (ACE), special test units (STU), bench maintenance equipment (BME), boilerplate and associate checkout equipment, and cabling systems.

7-33. Acceptance checkout equipment consists of permanently installed equipment located in control and computer rooms and carry-on (portable) equipment located near or onboard the spacecraft. Carry-on equipment is removed from the S/C prior to launch. ACE is computer-controlled equipment which provides the capability to check out spacecraft systems and to isolate malfunctions to a removable module. ACE also controls spacecraft systems servicing equipment.

7-34. Special test units provide the equipment to support the development of spacecraft systems and ACE systems. STU consists of manual checkout equipment and is required to operate and monitor the functional performance of the spacecraft systems.

7-35. Bench maintenance equipment is provided to perform verification and recertification, confirm defects, isolate malfunctions, perform repair verifications, perform calibration, and make adjustments on spacecraft systems, subsystems, and some components (to the lowest replaceable unit).

TEST FIXTURE (F-2)

SPACECRAFT 001

GROUND TEST ADAPTER

GROUND TEST STAND

SERVICE MODULE

A0026

WATER TANK

TEST STAND NO. 1
(TEST FIXTURE F-2
CURRENTLY IN OPERATION)

TEST FIXTURE (F-2)

PROPELLANT TRANSFER EQUIPMENT

CONTROL CENTER

SPACECRAFT 001
(SERVICE MODULE)

TEST STAND NO. 2
(FUTURE LOCATION OF
SPACECRAFT 001)

CRANE PAD

TO HOLDING PONDS

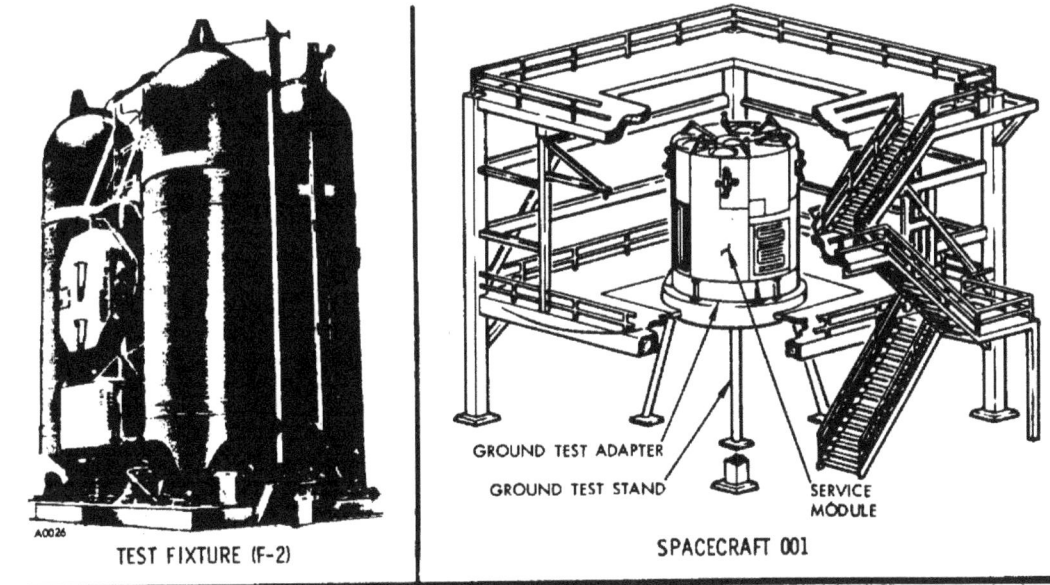

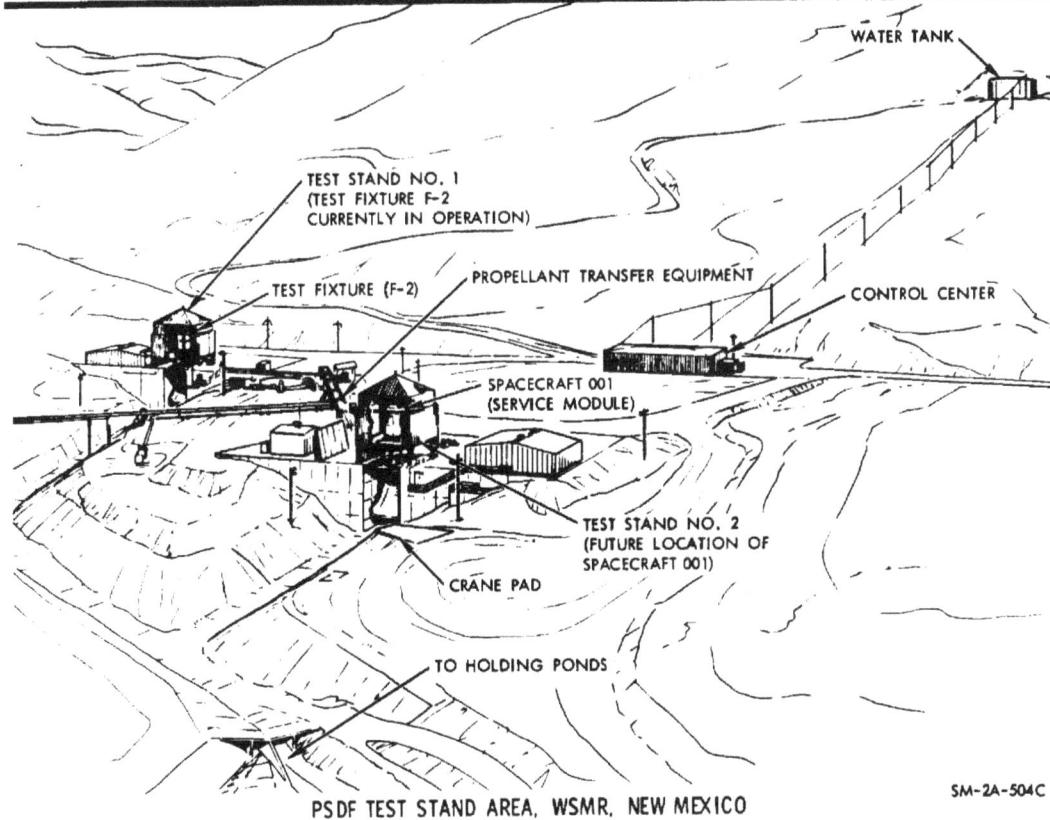

PSDF TEST STAND AREA, WSMR, NEW MEXICO

SM-2A-504C

Figure 7-6. Test Fixture (F-2) and Spacecraft 001 at Test Site

7-36. Boilerplate and associate checkout equipment consists of equipment that cannot be classified as ACE, STU, or BME. Boilerplate checkout equipment, such as the Apollo R&D instrumentation console and the onboard record checkout unit, are used to check out some boilerplates. Associate checkout equipment, such as the R-F checkout unit, supports the spacecraft systems when ACE or STU checkout equipment is being used, and equipment that cannot be isolated to one particular spacecraft system, such as the mobile recorder and spacecraft ground power supply and power distribution panel.

7-37. Cabling systems include that equipment necessary to provide electrical interconnection between various spacecraft vehicles, ground equipment, and test facilities, as required to provide an integrated electrical checkout station.

7-38. AUXILIARY EQUIPMENT. Auxiliary equipment consists of accessory equipment and special devices, such as substitute units, alignment equipment, and protective closures, which are not a part of the checkout, servicing, and handling equipment.

7-39. Substitute units, such as the launch escape tower substitute unit, service module substitute unit, and command module substitute unit, provide the intermodule interface required to support an interface compatibility checkout of the electrically unmated command module or service module.

7-40. Alignment equipment, such as the optical alignment set and the command module optical alignment support, provide the test fixtures required to accomplish alignment tasks.

7-41. Protective closures, both hard and soft, provide covering for flight equipment during transportation and storage.

7-42. SERVICING EQUIPMENT. Servicing equipment consists of fluid handling equipment necessary to support the spacecraft systems during ground operation, and fluid handling equipment necessary to permit the onboard loading of all liquids and gases required to operate the spacecraft systems. Servicing equipment provides additional capabilities such as flushing, purging, conditioning, vapor disposal necessary to support the basic servicing function, and decontaminating the spacecraft systems as required after detanking.

7-43. HANDLING EQUIPMENT. Handling equipment provides for lifting, transportation, weight and balance, alignment, access, protection, and support of the service module, command module, and launch escape assembly.

7-44. MISSIONS COMPLETED.

7-45. The following missions have been successfully completed.

7-46. BOILERPLATE 6.

7-47. Boilerplate 6, an unmanned, pad abort test vehicle, using the launch escape and pitch control motors as a launch vehicle, successfully completed its mission at the White Sands Missile Range, New Mexico, 7 November 1963. (See figure 7-7.) An abort command caused the launch escape and pitch control motors to ignite, lifting the command module from the launch pad adapter. At an approximate altitude of 5000 feet, the launch escape assembly and forward heat shield were separated from the C/M, and the tower jettison motor ignited, propelling the launch escape assembly and forward heat shield clear of the C/M trajectory. The earth landing system was initiated to accomplish drogue parachute deployment, and release and deployment of three pilot parachutes which, in turn, deployed the three main parachutes, slowing the C/M to a safe landing speed (approximately 25 feet per second). Boilerplate 6 is being refurbished for further parachute recovery system tests and will be designated boilerplate 6A.

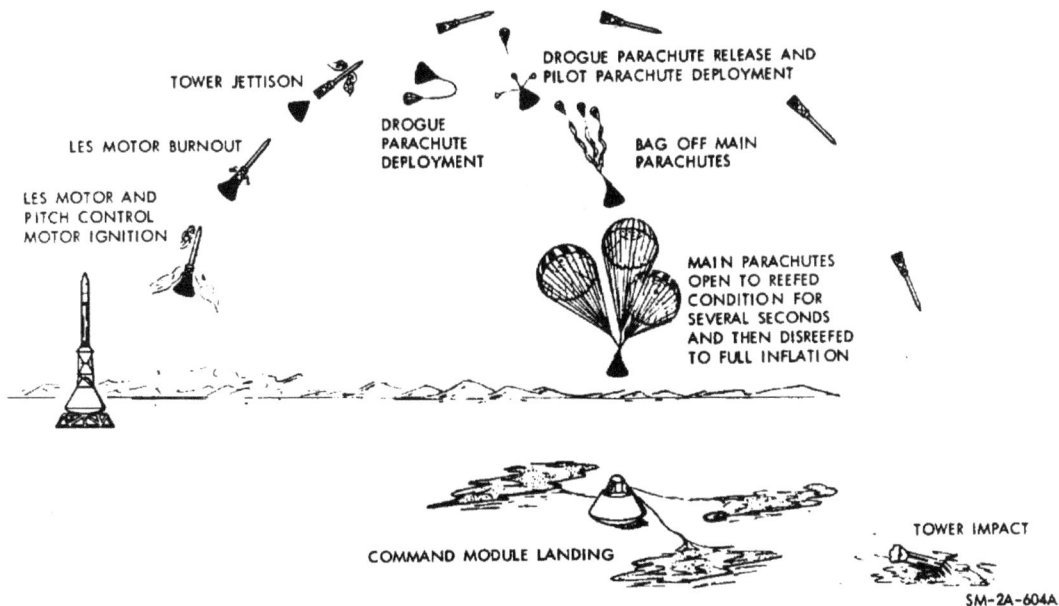

Figure 7-7. Boilerplate 6 Mission Profile

7-48. BOILERPLATE 12.

7-49. Boilerplate 12, an unmanned, transonic abort test vehicle, using a Little Joe II booster as a launch vehicle, successfully completed its mission at WSMR, New Mexico, 13 May 1964. (See figure 7-8.) This was the first full-scale test flight of the launch escape system in the transonic speed range. The Little Joe II boosted the boilerplate command-service modules to an approximate altitude of 21,000 feet, where an abort command caused separation of the C/M from the S/M and ignition of the launch escape and pitch control motors. The launch escape assembly propelled the command module away from the S/M and launch vehicle to an approximate altitude of 28,000 feet. The tower was separated from the C/M and the tower jettison motor ignited, carrying the launch escape assembly and forward compartment heat shield away from the trajectory of the C/M. The earth landing system was then initiated to accomplish drogue parachute deployment and release, and deployment of three pilot parachutes which, in turn, deployed the three main parachutes. One main parachute did not inflate fully and was separated from the C/M; however, the boilerplate 12 command module landed upright and undamaged.

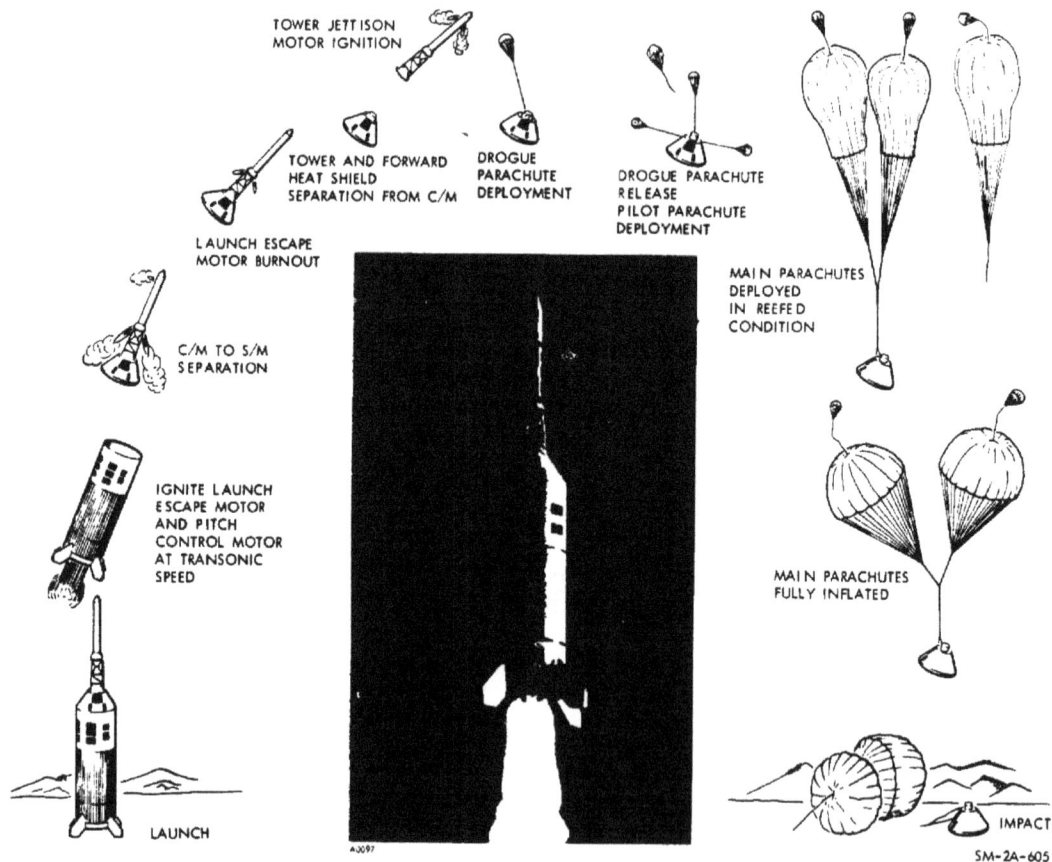

Figure 7-8. Boilerplate 12 Mission Profile

7-50. BOILERPLATE 13.

7-51. Boilerplate 13, an unmanned, launch environment test vehicle, using a Saturn I as a launch vehicle, was successfully launched into orbit from Kennedy Space Center, Florida, 28 May 1964. (See figure 7-9.) This was the first test flight to qualify the Saturn I launch vehicle and to demonstrate the compatibility of the spacecraft and launch vehicle. All test objectives were met; design parameters and conclusions about the flight, based on ground research, were as predicted. Orbits of the CSM and second-stage booster ranged from 110 to 140 miles above the earth surface and continued until 31 May 1964. Upon entry into earth atmosphere, the test vehicle disintegrated, as no provisions were made for recovery.

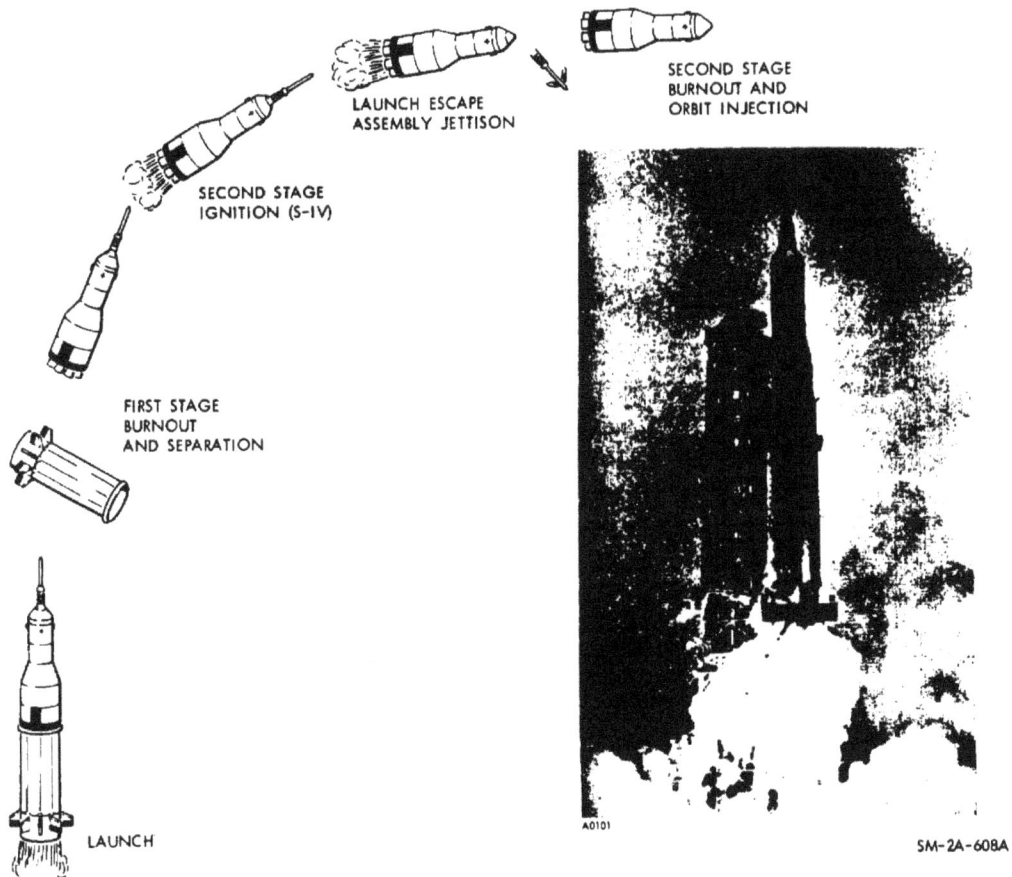

SECOND STAGE
BURNOUT AND
ORBIT INJECTION

LAUNCH ESCAPE
ASSEMBLY JETTISON

SECOND STAGE
IGNITION (S-IV)

FIRST STAGE
BURNOUT
AND SEPARATION

LAUNCH

A0101

SM-2A-608A

Figure 7-9. Boilerplate 13 Mission Profile

7-52. BOILERPLATE 15.

7-53. Boilerplate 15, an unmanned, launch environment test vehicle, using a Saturn I as a launch vehicle, was successfully launched into orbit from Kennedy Space Center, Florida, 18 September 1964. (See figure 7-10.) This was the second successful test flight to qualify the Saturn I launch vehicle and to demonstrate compatibility of the spacecraft and launch vehicle. An alternate mode of jettisoning the launch escape assembly was also demonstrated. Orbits of the CSM and second-stage booster ranged from 115 to 141 miles above the earth's surface and continued until 22 September 1964. No provisions were made for recovering the test vehicle upon entry into the atmosphere of the earth.

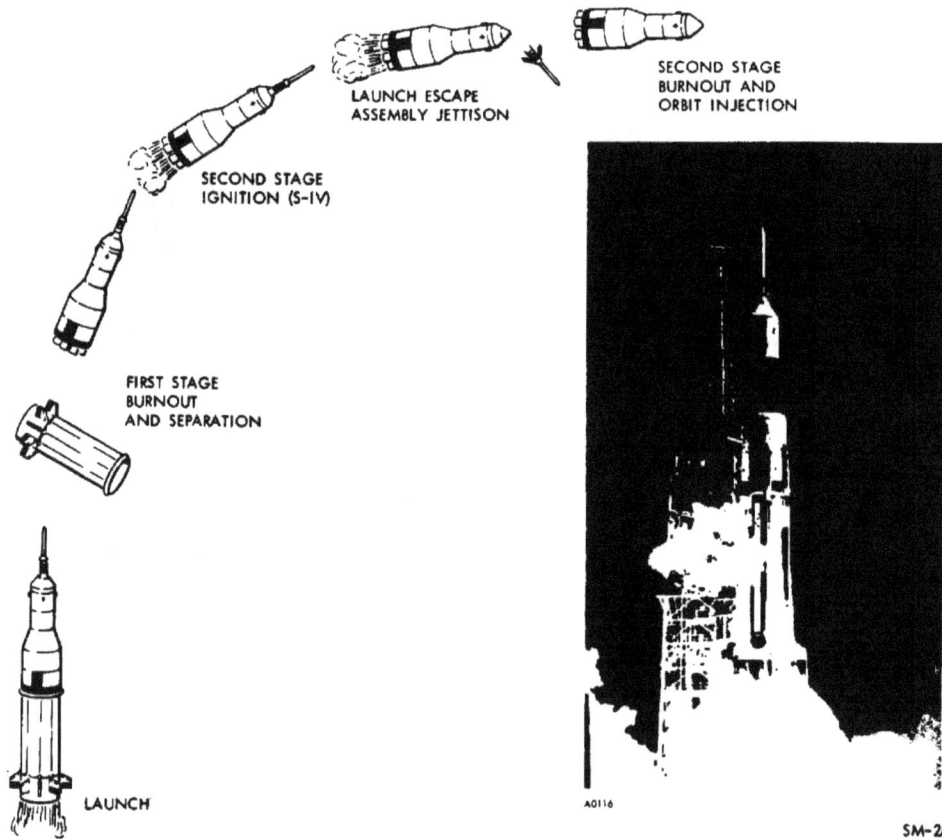

Figure 7-10. Boilerplate 15 Mission Profile

7-54. BOILERPLATE 23.

7-55. Boilerplate 23, an unmanned, abort test vehicle, using a Little Joe II booster as a launch vehicle, successfully completed its mission at WSMR, New Mexico, 8 December 1964. (See figure 7-11.) At approximately 32,000 feet a radio command signalled the launch vehicle control system to produce a pitch-up maneuver, simulating an abort condition. An abort command was initiated at an approximate altitude of 35,000 feet. Upon receipt of the abort signal, the C/M-S/M separated, and the launch escape and pitch control rocket motors ignited, to carry the C/M away from its launch vehicle. Eleven seconds after abort was initiated, the canards deployed, turning the C/M around and stabilizing it in a blunt-end forward attitude. At approximately 25,000 feet, the launch escape assembly separated from the C/M and the tower jettison motor ignited, carrying the launch escape assembly, boost protective cover, and forward heat shield away from the C/M. The earth landing system was initiated, accomplishing: drogue parachute (2) deployment in a reefed condition; drogue parachutes disreefed (at approximately 11,000 feet) slowing the speed and oscillation of the C/M for main parachute deployment, drogue parachute release, pilot parachute (3) deployment which, deployed the three main parachutes in a reefed condition; main parachutes disreefed lowering C/M to ground at a safe landing speed (approximately 25 feet per second). Boilerplate 23 is being refurbished for further C/M pad abort tests and will be designated boilerplate 23A.

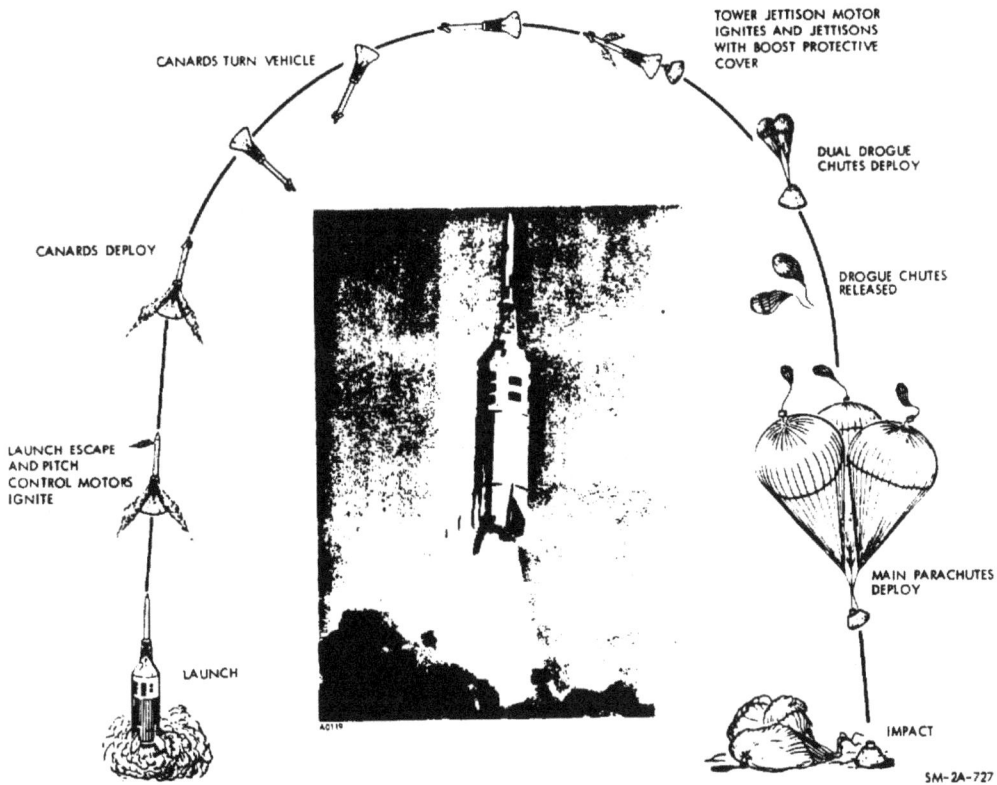

Figure 7-11. Boilerplate 23 Mission Profile

7-56. BOILERPLATE 16.

7-57. Boilerplate 16, an unmanned, micrometeoroid experiment test vehicle, using a Saturn I as a launch vehicle, was successfully launched into orbit from Kennedy Space Center, Florida, 16 February 1965. (See figure 7-12.) Once the orbit was attained, the CSM was jettisoned from the second stage (S-IV) by using the LES, and two large (NASA-installed) panels, unfolded. The panels and associated electronics, installed in the S-IV, are used to detect micrometeoroid particles and transmit the information to ground stations. The orbit of the test vehicle ranges from 308 to 462 miles above the earth. No provisions have been made for recovering the test vehicle upon entry into the atmosphere of the earth.

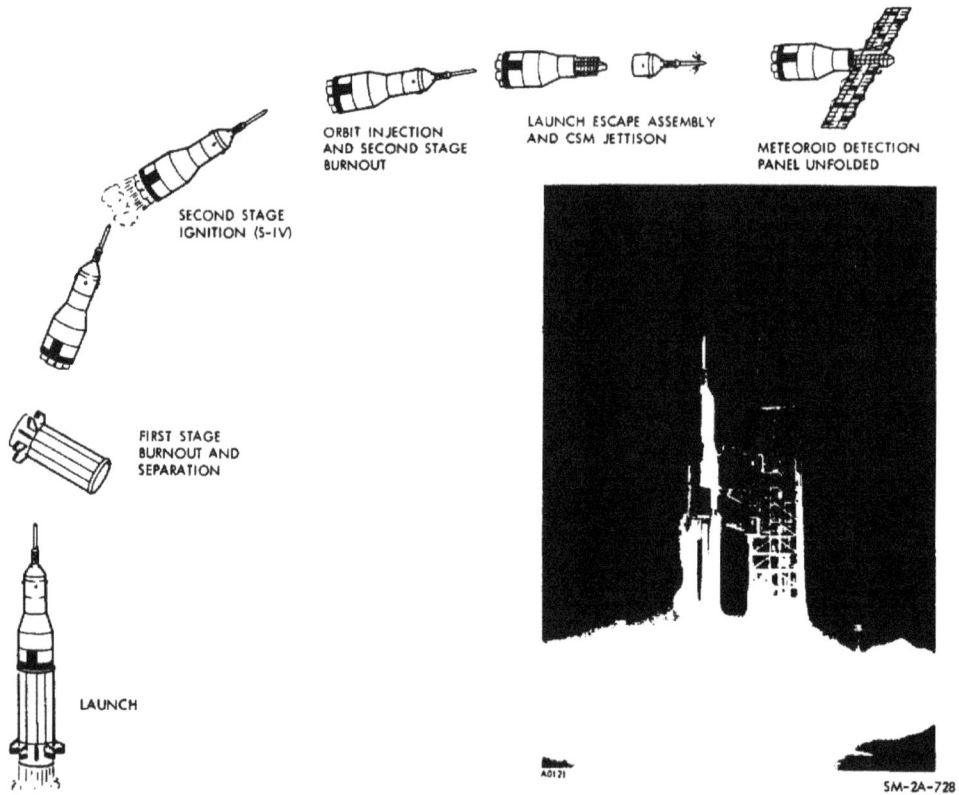

Figure 7-12. Boilerplate 16 Mission Profile

7-58. BOILERPLATE 22.

7-59. Boilerplate 22, an unmanned abort test vehicle using a Little Joe II booster as a launch vehicle, was partially successful in completing its mission at WSMR, New Mexico, 19 May 1965. (See figure 7-13). Although a high-altitude abort was planned, the boost vehicle malfunctioned causing a premature low-altitude abort; however, the Apollo systems functioned perfectly. An abort command was initiated due to the malfunctioning boost vehicle. Upon receipt of the abort signal, the C/M-S/M separated, and the launch escape and pitch control rocket motors ignited carrying the C/M away from the launch vehicle debris. The earth landing system was initiated lowering the C/M safely to the ground.

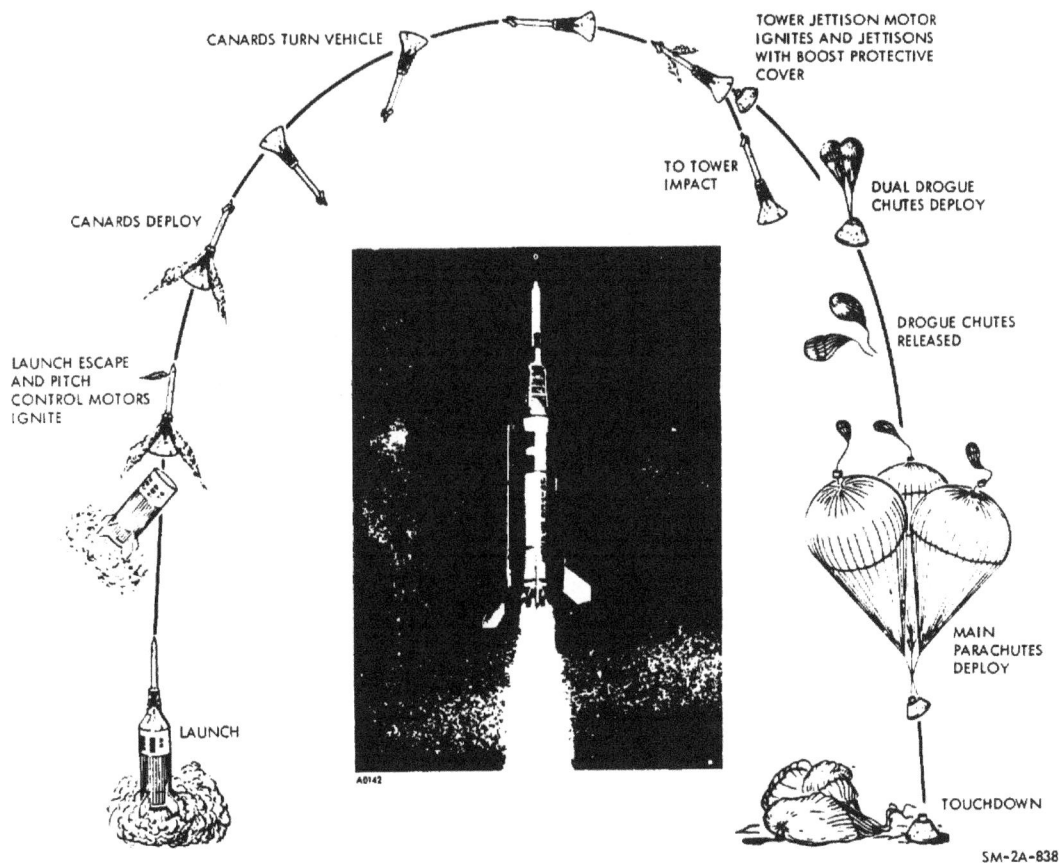

Figure 7-13. Boilerplate 22 Mission Profile

7-60.   BOILERPLATE 26.

7-61.   Boilerplate 26, the second unmanned micrometeoroid experiment test vehicle, was successfully launched into orbit from Kennedy Space Center, Florida, 25 May 1965.   The launch vehicle used was a Saturn I.   Using the LES, the CSM was jettisoned from the second stage (S-IV) upon reaching the planned orbit.   (See figure 7-14.)   Two large (NASA-installed) panels installed in the S-IV, unfolded, to be used with associated electronics to detect micrometeoroid particles and transmit the information to ground stations.   No provisions have been made for recovering the test vehicle upon entry into the atmosphere of the earth.

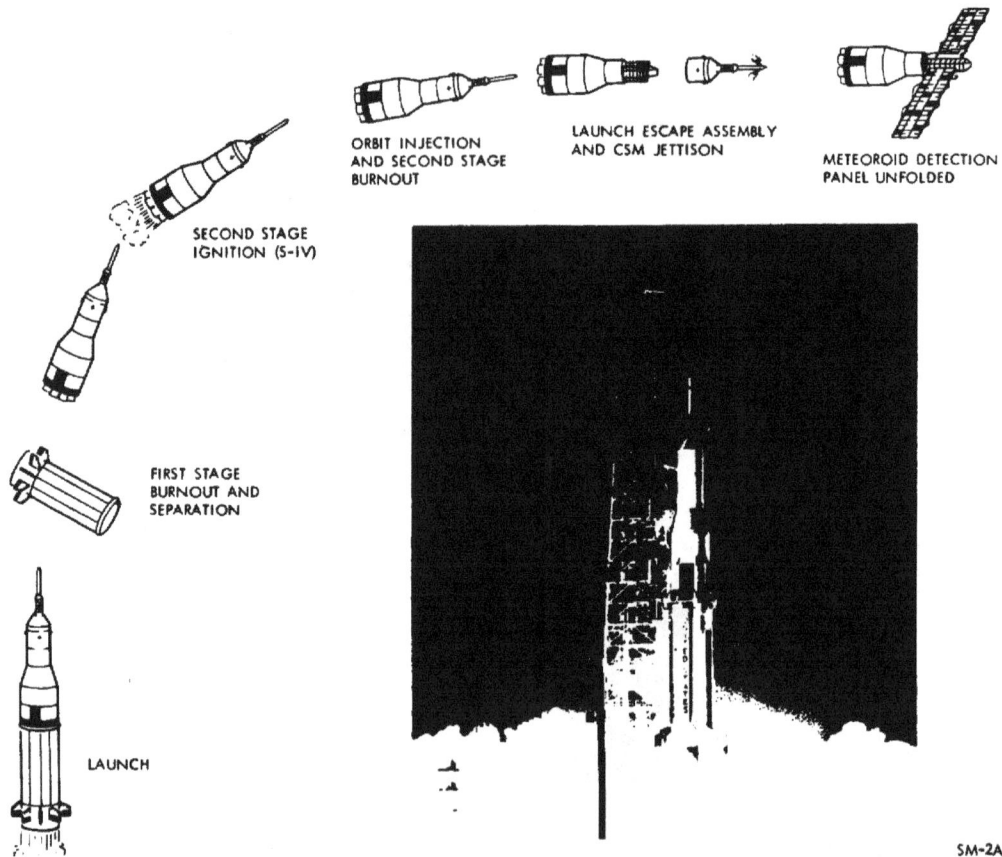

Figure 7-14.   Boilerplate 26 Mission Profile

7-62. BOILERPLATE 23A.

7-63. Boilerplate 23A, another pad abort test vehicle, successfully completed its mission at the White Sands Missile Range, New Mexico, 29 June 1965. (See figure 7-15.) This was the second test of the launch escape systems ability to lift the C/M off the pad. The test simulated an emergency abort which might occur while the C/M was still on the launch pad atop a Saturn launch vehicle. An abort command was initiated causing the launch escape and pitch control motors to ignite, the C/M to separate from the S/M, and lifting the C/M from the launch pad adpater. Improvements incorporated in boilerplate 23A which were not included on the first pad abort test vehicle were: canard surfaces, boost protective cover, a jettisonable forward compartment heat shield, and reefed dual drogue parachutes. All systems worked as predicted and the C/M was lowered to the ground safely.

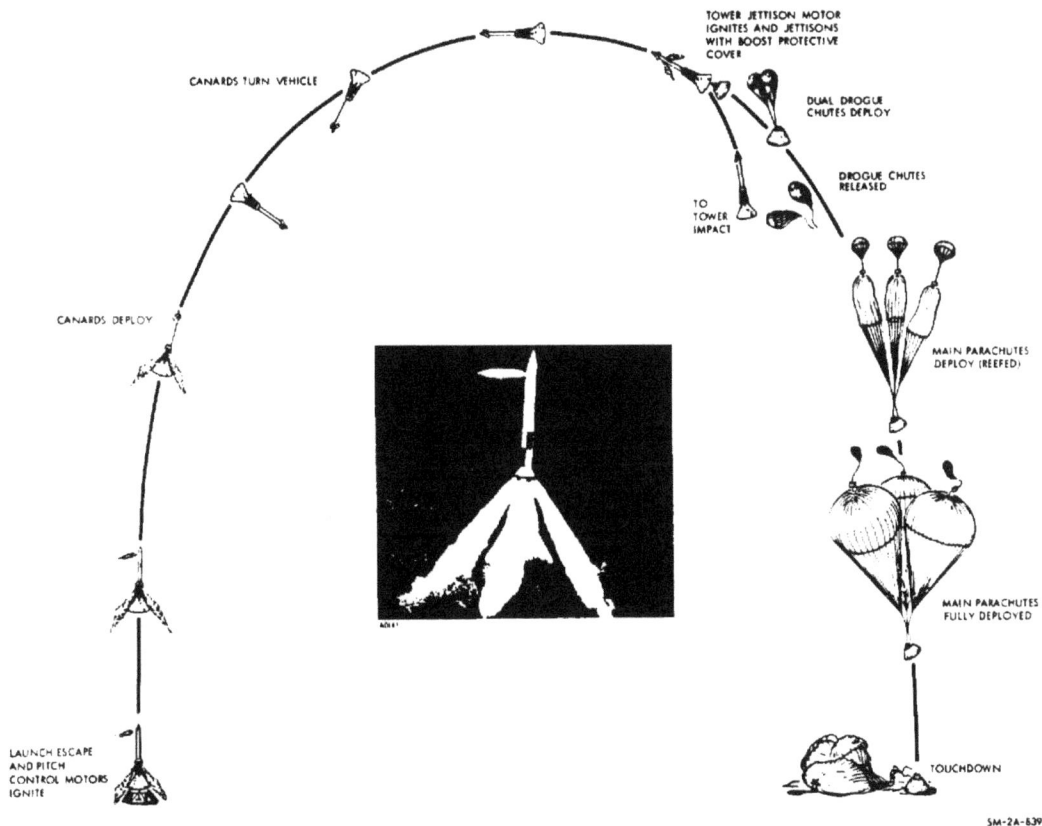

Figure 7-15. Boilerplate 23A Mission Profile

7-64. BOILERPLATE 9A.

7-65. Placing the third micrometeoroid experiment test vehicle successfully into orbit, was accomplished 30 July 1965, from Kennedy Space Center, Florida. Boilerplate 9A was used as a cover for the folded test panels, and a Saturn I with an S-IV second stage was used as the launch vehicle. (See figure 7-16.) After reaching its orbit, the CSM was jettisoned using the LES, and the large (NASA-installed) panels unfolded from the second stage (S-IV), the same as the two previous micrometeoroid test vehicles. Micrometeoroid particles are detected when they strike the panels, and the information is transmitted to ground stations by way of electronics installed in the S-IV. The test vehicle will not be recovered upon entry into the atmosphere of the earth.

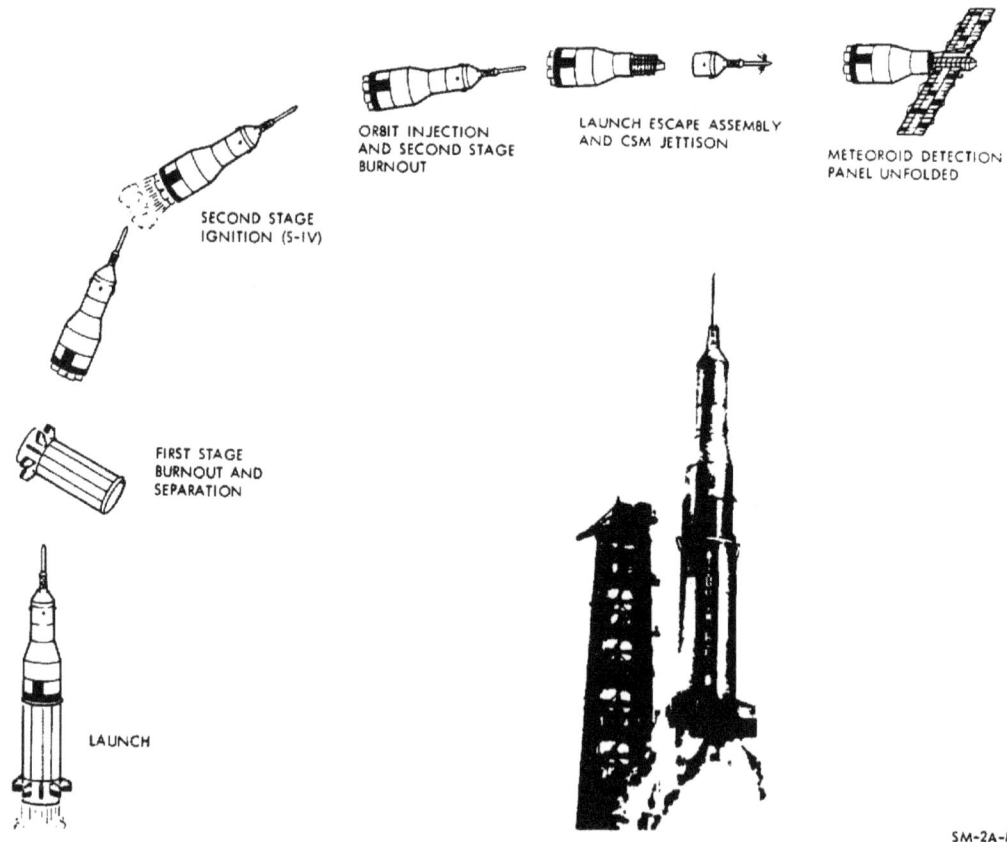

ORBIT INJECTION
AND SECOND STAGE
BURNOUT

LAUNCH ESCAPE ASSEMBLY
AND CSM JETTISON

METEOROID DETECTION
PANEL UNFOLDED

SECOND STAGE
IGNITION (S-IV)

FIRST STAGE
BURNOUT AND
SEPARATION

LAUNCH

SM-2A-841

Figure 7-16. Boilerplate 9A Mission Profile

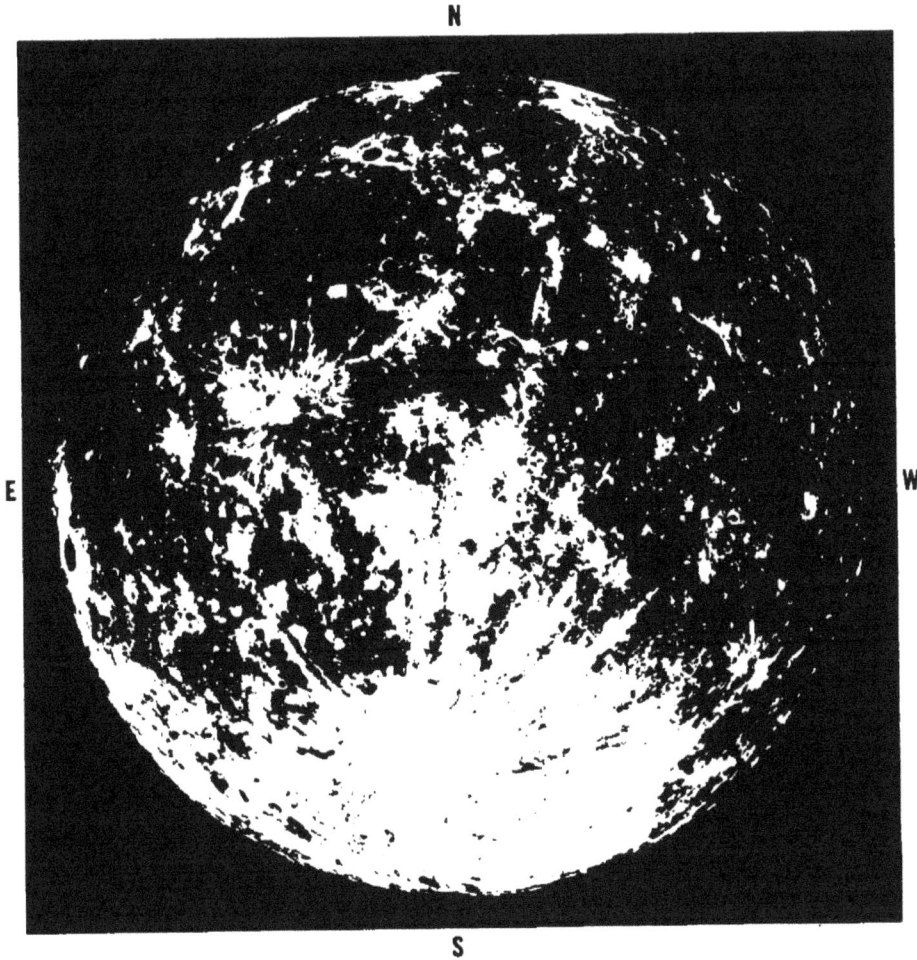

LUNAR DATA

DISTANCE FROM EARTH – 253,000 MILES (MAX.)

DIAMETER   2160 MILES

TEMPERATURE
   SUN AT ZENITH   214°F (101°C)
   NIGHT   APPROX. – 250°F (–157°C)

SM-2A-878

LUNAR LANDING MISSION

8-1. GENERAL.

8-2. The culmination of the Apollo program will be the lunar landing mission.
This mission will produce the first extraterrestrial-manned exploration of the
moon. This section contains a sequential presentation of the major events of
the lunar landing mission. Text and illustrations within this section (figures 8-1 through 8-23) provide general information concerning the operations
involved in the lunar landing mission. Text and corresponding illustrations
are connected by the use of common titles.

8-3. KENNEDY SPACE CENTER.

8-4. The lunar landing mission will originate from the Kennedy Space Center
(KSC). Facilities have been constructed at KSC to handle space-exploration
vehicles and associated equipment. These special facilities provide capability to handle the large vehicles and components within precise parameters.
Figure 8-1 shows launch pad B of complex 39 in the foreground, the remote
blockhouse to the left, the 500-foot-plus vehicle assembly building (VAB) in
the background, and the interconnecting crawlerway.

8-5. The component assemblies of the Apollo spacecraft and the Saturn V
launch vehicle will be transported to KSC for final assembly tests. The
spacecraft and launch vehicle will be assembled (stacked) on the launch umbilical tower (LUT) platform in the vehicle assembly building. The launch
umbilical tower platform is mounted on the crawler-transporter. After
assembly is completed, interface and systems tests will be made.

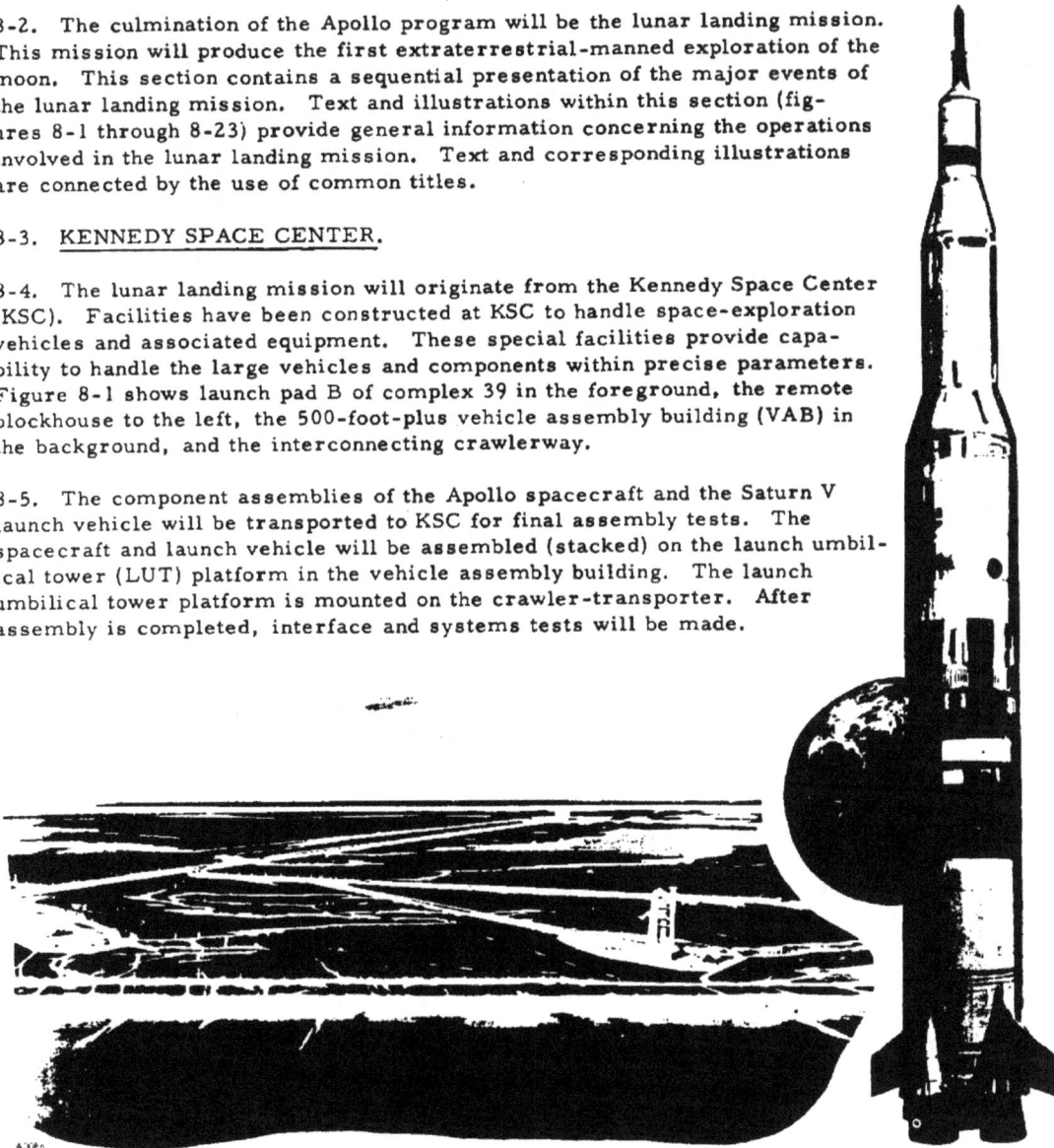

Figure 8-1. Kennedy Space Center (KSC)

Figure 8-2.   Transportation to Launch Pad

8-6.   TRANSPORTATION TO LAUNCH PAD.

8.7 After operations are concluded in the vehicle assembly building, the assembled LUT, spacecraft, and launch vehicle will be transported to launch complex 39. Transportation of the launch umbilical tower, spacecraft, and launch vehicle will be provided by the crawler-transporter. The crawler-transporter will carry this load 3.5 miles to launch pad A and 4.7 miles to launch pad B on a specially constructed crawlerway. The crawlerway is a pair of parallel roadways which can support a load in excess of 18-million pounds. The crawler-transporter will proceed at a rate of approximately 1 mile per hour.

8-8. LAUNCH PAD.

8-9. Upon arrival at the launch pad, the crawler-transporter will lower the launch umbilical tower, platform, spacecraft, and launch vehicle onto steel foundations. The crawler-transporter will move a mobile service structure onto the pad next to the spacecraft. The MSS provides facilities for pyrotechnic arming and fueling operations. When the MSS is no longer needed, the crawler-transporter and the MSS will be removed from the launch area.

A0088   SM-2A-509A

Figure 8-3. Launch Pad

A0089

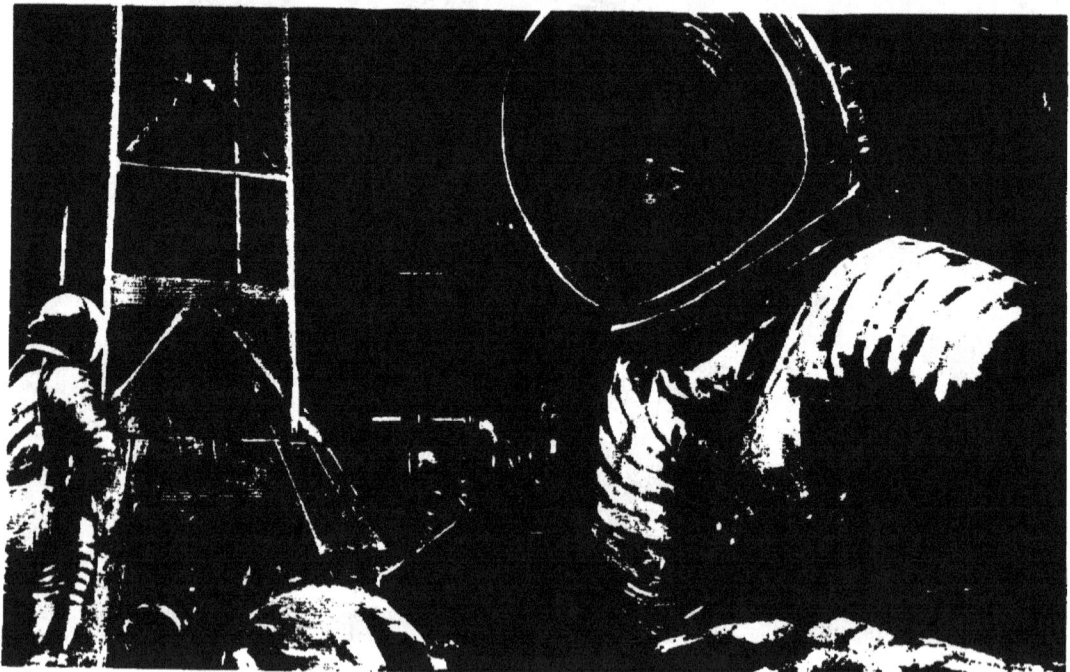

A0090

SM-2A-510A

Figure 8-4. Countdown

8-10. <u>COUNTDOWN.</u>

8-11. The final prelaunch countdown sequence begins upon final positioning of the launch umbilical tower, spacecraft, and launch vehicle on the launch pad. Ordnance components required on the spacecraft are armed, utilizing the MSS. Appropriate protective devices are installed at the time of ordnance installation to prevent inadvertent operational arming and firing of the pyrotechnics, and to provide maximum safety for the spacecraft checkout crew and launch area ground personnel.

8-12. The prelaunch countdown follows a programed sequence which is directed by, and under the control of, the launch control director. This sequence establishes the order of required operational checkout of the spacecraft systems and of the servicing and loading of consumable gases, fuels, and supplies.

8-13. The countdown sequence consists essentially of the activation, or simulated activation, and verification checks of the spacecraft operational systems as follows:

- Removal of ground support equipment (GSE)
- Leak checks
- Battery activation
- Final arming of ordnance devices
- Removal of ordnance shorting devices
- Loading of fuels: helium, liquid hydrogen, and liquid oxygen
- Fuel cell activation
- Entry of mission flight crew into command module
- Closing of command module crew hatch
- Installation of boost protective cover hatch access cover
- Command module crew cabin leak check
- Purging the command module cabin with 100-percent oxygen
- Final confidence checks of the spacecraft systems by the crew
- Final arming of the launch escape system
- Ground-to-spacecraft umbilical disconnect.

8-14. Upon completion of final checks, the ground-to-spacecraft umbilical cables are disconnected and the launch tower support arms are retracted. Final decision and approval to launch is verified by the launch control center and the spacecraft crew.

Figure 8-5. Lift-Off

8-15. **LIFT-OFF.**

8-16. Upon launch, the Saturn V first-stage (S-IC) engines are ignited by the launch control center. The center engine ignites first, followed by the ignition of the four outer engines. The launch pad hold-down devices release after initial operational thrust is sufficient. The launch control center and the spacecraft crew will continuously monitor the initial operational ascent attitude parameters.

8-17. FIRST-STAGE SEPARATION.

8-18. The launch vehicle guidance system initiates roll of the spacecraft to the required launch azimuth. The first-stage pitch programer initiates the required pitchover of the spacecraft. Voice communication between the crew and manned space-flight network (MSFN) is maintained through the critical maximum dynamic flight conditions and throughout the ascent phase. The cutoff of the first-stage engines is followed by ignition of the second-stage ullage rockets. The first-stage retrorocket then separates the first stage from the second stage.

SM-2A-512A

Figure 8-6. First-Stage Separation

8-19. SECOND-STAGE EVENTS.

8-20. Second-stage (S-II) engine ignition occurs nominally two seconds after cutoff of the first-stage engines at approximately 200,000 feet. The flight trajectory of the spacecraft is controlled by the third stage (S-IVB) inertial guidance system. The launch escape system is operationally jettisoned at approximately 320,000 feet altitude. The second-stage engines are cut off at an altitude of approximately 600,000 feet. The sequential ignition of the third-stage (S-IVB) ullage rockets, second-stage retrorockets, and third-stage engines effects separation of the second stage. The third-stage engines provide the thrust required to place the spacecraft in earth orbit. The third-stage guidance control system cuts off the third-stage engines after the programed orbit conditions have been attained.

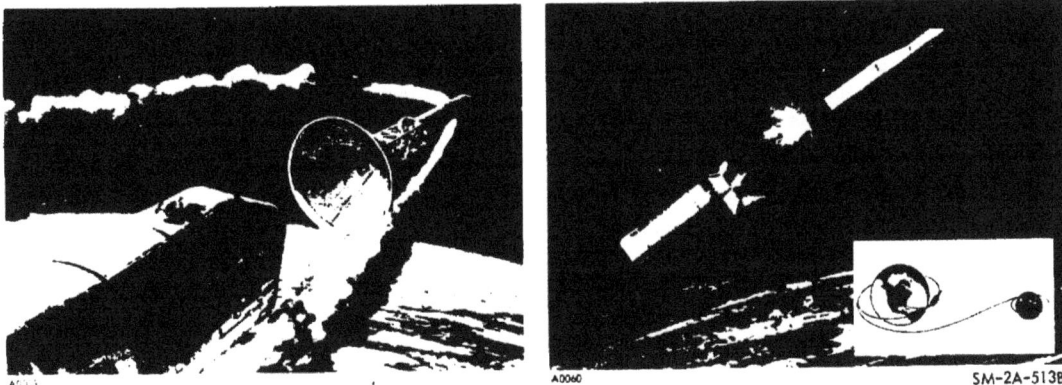

SM-2A-513B

Figure 8-7. Second-Stage Events

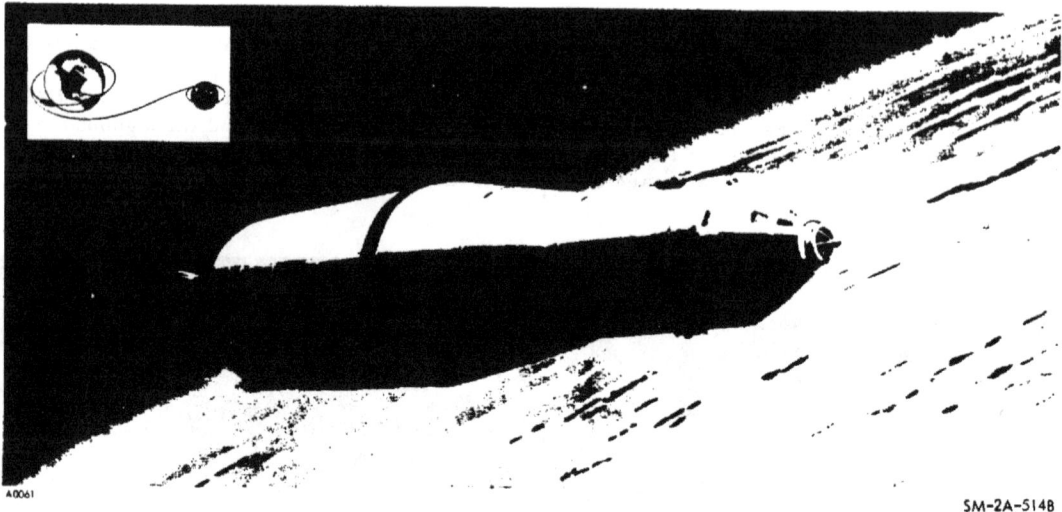

A0061

SM-2A-5148

Figure 8-8. Earth Orbit

8-21. __EARTH ORBIT.__

8-22. The spacecraft and third stage are to orbit the earth, no more than three times, at an approximate altitude of 100 nautical miles. During this period, the orbital parameters are determined by manned space-flight network; then verified by landmark navigational sightings and the spacecraft crew. This determines the required velocity increment and trajectory for translunar injection.

8-23. The crew will perform a biomedical and safety equipment check. Sequence checks will be made of the environmental control system, communications and instrumentation system, service propulsion system, service module reaction control system, electrical power system, guidance and navigation system, stabilization and control system, and crew equipment system.

8-24. Translunar injection parameters are determined onboard the spacecraft by sequential landmark navigational sightings (using the scanning telescope) and by the Apollo guidance computer. Trajectory and star-tracking data computations are made by the Apollo guidance computer. The inertial measurement unit is fine-aligned for the translunar injection monitor, using the Apollo guidance computer. The center-of-gravity offset angles are set into the service propulsion system gimbal position display. The ΔV program, time, and direction vector are set into the Apollo guidance computer. The stabilization and control system is prepared for the ΔV maneuver, including minimum deadband hold control and monitor mode. Finally, the third-stage reaction control system is prepared for the ΔV translunar injection.

8-25. Verification of "go" conditions for translunar injection will be confirmed by the spacecraft crew and the MSFN. The third-stage countdown and ignition sequence is performed with the spacecraft in the required translunar injection attitude.

A0062   SM-2A-515A

Figure 8-9.   Translunar Injection

8-26.   TRANSLUNAR INJECTION.

8-27.   The translunar injection phase begins with the third-stage ullage rockett ignition. The third-stage propulsion system is operated to provide sufficient thrust to place the spacecraft in a translunar "free-return" trajectory in accordance with the ΔV magnitude, time duration, and thrust vector previously established and operationally programed onboard the spacecraft by the crew, and confirmed by the MSFN.

8-28.   The third-stage instrument unit provides operational guidance control for the translunar injection with the Apollo guidance and navigation system capable of backup control, if necessary.   The third-stage engines operate for the predetermined time, nominally 5 minutes.   The crew monitors the emergency detection system and spacecraft attitude control displays.   The spacecraft guidance and navigation system monitors the programed injection maneuver.

8-29.   INITIAL TRANSLUNAR COAST.

8-30.   Following translunar injection, the MSFN will determine the spacecraft trajectory and verify it with an onboard determination performed by the crew.   The operational controls are then set for an initial coast phase.   An onboard systems check is then made of all crew equipment, electrical power system, environmental control system, service module reaction control system, and the service propulsion system.   The status of these systems is communicated to MSFN.

8-31.   The spacecraft body-mounted attitude gyros are aligned, using the third-stage stable platform as a reference, and the flight director attitude indicator is set preparatory to initiating transposition of the lunar module.   Confirmation of conditions for initiating transposition of the lunar module is made with MSFN.

A0062   SM-2A-516B

Figure 8-10.   Initial Translunar Coast

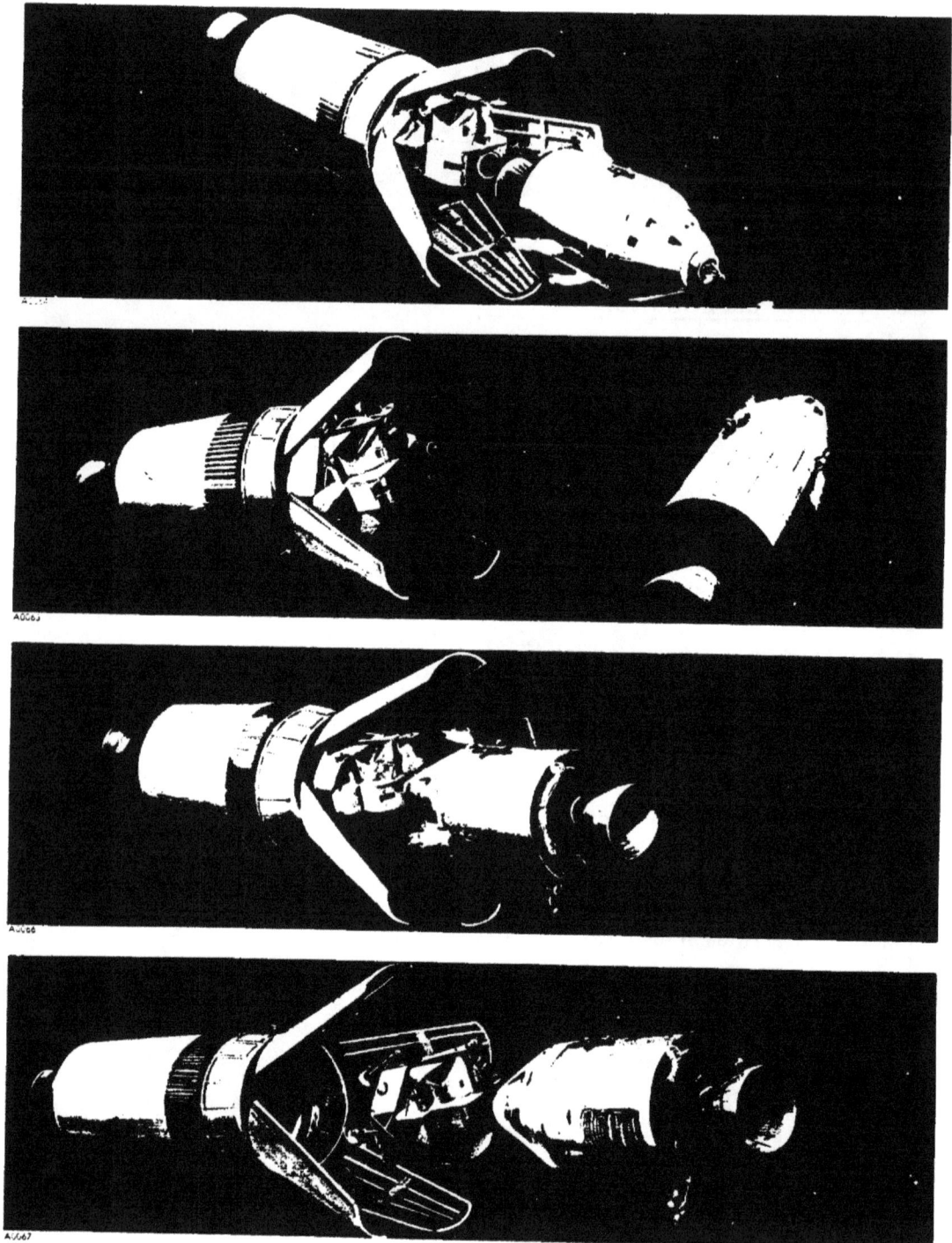

Figure 8-11. Spacecraft Transposition and Docking

SM-2A-517E

8-32. SPACECRAFT TRANSPOSITION AND DOCKING.

8-33. Transposition and docking of the spacecraft consists essentially of separating and translating the command/service module from the spacecraft-LM-adapter (SLA), third stage, pitching the command/service module 180 degrees, and translating the command/service module back to the lunar module to join the lunar module to the command module. The S-IVB guidance system stabilizes the SLA and third stage, during the transposition operations which precede docking. Upon completion of docking, the third stage is jettisoned. The entire maneuver will normally be completed within one hour after translunar injection. Necessary precautions will be observed by the crew during the time the spacecraft passes through the Van Allen belts.

8-34. The spacecraft will be oriented within communication constraints to provide the most desirable background lighting conditions for the transposition of the spacecraft. The third stage is stabilized in an attitude-hold mode. The adapter is pyrotechnically separated from the command/service module, which is then translated approximately 50 feet ahead of the lunar module and third stage, using the service module reaction control system engines. The command/service module is then rotated 180 degrees in pitch, using the service module reaction control engines. The docking attitude of the command/service module for the SLA, third stage will be established and maintained, using the service module reaction control system engines.

8-35. The command/service module will be translated toward the SLA, third stage, with minimum closing velocity, so that the command/service module docking probe engages within the drogue mechanism on the lunar module. A mechanical latching assembly secures the lunar module to the command/service module. The command/service module, with the lunar module attached, then separates and translates away from the third stage.

A0C68

A0069

Figure 8-12.   Final Translunar Coast

SM-2A-518C

8-36. FINAL TRANSLUNAR COAST.

8-37. The final translunar coast phase begins with the ignition of the service module reaction control system to separate the spacecraft from the third stage, and ends just prior to lunar orbit insertion. The primary operations occurring during this phase consist of spacecraft systems checkout, trajectory verifications, and preparation for lunar orbit insertion. Midcourse ΔV corrections, navigational sightings, and inertial measurement-unit alignments are to be made.

8-38. The spacecraft guidance and navigation system computes the trajectory of the space-craft (in conjunction with navigational sightings). The delta increment required is deter-mined by MSFN, and confirmation of the trajectory and velocity increment values is made with the Apollo guidance computer. Midcourse incremental velocity corrections will be made when required.

8-39. The attitude of the spacecraft will be constrained at times because of operational temperature control restrictions. At least one astronaut will be in his space suit at all times. A crew work-rest cycle will be established and followed during this phase. The capability to initiate an abort at any time during this space will be provided.

8-40. In preparation for lunar orbit insertion, the spacecraft attitude, lunar orbit insertion velocity increment, and the time to initiate the service propulsion system thrust required to achieve the desired orbit around the moon are determined by trajectory data from MSFN and from the guidance and navigation system.

Figure 8-13. Lunar Orbit Insertion

8-41. LUNAR ORBIT INSERTION.

8-42. This phase begins with the spacecraft properly oriented for lunar orbit insertion and ends with the cutoff of the service propulsion system as the spacecraft is inserted into a lunar orbit.

8-43. Navigation sightings will be made using the Apollo guidance computer, inertial measurement unit, and scanning telescope. The MSFN determines the lunar orbit insertion trajectory and star catalog data, ΔV correction parameters, and the lunar orbit insertion parameters. These determinations are confirmed using the Apollo guidance computer. The reaction control system ignition provides a translation impulse and roll-control operation. The guidance and navigation system initiates and controls the programed insertion maneuvers and service propulsion system.

8-44. The retrograde impulse required to establish the lunar orbit occurs near the minimum altitude of the lunar approach trajectory. The point of this altitude is almost directly behind the moon with respect to the earth. The total velocity increment required to achieve the desired orbit around the moon, including any necessary plane changes, is applied during this phase.

8-45. The initial lunar orbit coast phase begins with cutoff of the service propulsion engine as the spacecraft is inserted into lunar orbit and ends with activation of the lunar module reaction control system to effect separation from the spacecraft.

8-46. Following lunar orbit insertion, the crew will transmit trajectory data and information to MSFN. The orbit ephemeris about the moon will be determined as accurately as possible, using the spacecraft guidance system and MSFN. A confirming checkout of the lunar module guidance system is also made prior to separation from the spacecraft.

8-47. The CSM systems will be capable of operation at their nominal design performance level for a mission of approximately 11 days. A single crewmember can control the spacecraft in lunar orbit for several days. Communication capability will be provided between the CSM, manned space-flight network, and the lunar module when separated and within line-of-sight. The CSM and lunar module separation and docking operations will not be restricted by natural illumination conditions.

8-48. Observations and calculations will be made of the preselected landing site from the CSM, to determine if the area location is satisfactory or if an alternate landing area should be selected. Detailed surveillance of the landing area is to be made from the lunar module prior to landing.

8-49. Lunar orbit trajectory verification requires fine-alignment of the inertial measurement unit, and a related series of navigational sightings will be made of known lunar surface areas and reference stars, using the scanning telescope, sextant, and Apollo guidance computer. Parameters for lunar orbit and transearth injection will be determined by MSFN and confirmed by the Apollo guidance computer; but the LM descent trajectory will be calculated using the LM guidance, navigation, and control system.

8-50. Upon final confirmation of these parameters, the commander and the pilot will transfer from the CSM to the lunar module. The lunar module electrical power system, environmental control system, communication system, guidance, navigation, and control system, reaction control system, and ascent and descent engine systems will be checked out and the landing gear extended. A check will be made of emergency procedures and corresponding spacecraft systems. The operational capability of the air lock will be verified. Initial operational information will be synchronized between the CSM and lunar module.

8-51. The CSM will be aligned and held in the required attitude for separation. The lunar module guidance computer will be programed for the transfer trajectory when the CSM orbit is determined accurately. The orbit will not be disturbed, unless an emergency requirement prevails, or until the docking phase is complete. Emergency or additional data may require that the CSM lunar orbit be updated as necessary by the remaining crewmember.

8-52. Actual separation of the lunar module from the CSM is effected by a propulsion thrust from the lunar module reaction control system. After a specified time, an equivalent impulse is applied in the opposite direction so that the relative velocity between the lunar module and the CSM will be zero during the final checkout of the lunar module. Final checkout is accomplished with the lunar module in free flight, but relatively close to the CSM in case immediate docking is required.

Figure 8-14. Lunar Landing

SM-2A-520D

8-53. LUNAR LANDING.

8-54. The lunar landing operations phase begins with inital lunar module separation from the CSM and ends with touchdown on the surface of the moon.

8-55. The essential lunar module operations which occur are attitude control, incremental velocity control, and time to fire, computed by the lunar module guidance, navigation, and control system and verified by the CSM guidance and navigation system. Lunar module insertion into a descent orbit is accomplished by reaction control system ullage acceleration and ignition of the descent engine with the thrust level and burn-time automatically controlled by the lunar module guidance navigation, and control system. The lunar module separates from the CSM during this maneuver, and communication with manned space-flight network at time of insertion cannot be accomplished, since it occurs on the far side of the moon.

3-56. Descent trajectory determination will be made using data from the LM rendezvous radar tracking the CSM. The lunar module will coast in a descent transfer orbit following descent engine cutoff, and close observation of the proposed lunar landing site will be made for final approval. The descent engine will be re-ignited prior to reaching the low point in the orbit and sustained thrust initated for the descent maneuver. The thrust and attitude are controlled by the guidance, navigation and control system by comparison of the actual and planned landing tracks.

8-57. The translational and radial velocities will be reduced to small values and the descent engine cut off at a specified altitude above the lunar surface. The commander will control the descent within present limits. Terminal descent and touchdown will be made by manual control and use of the landing radar. Confirmation of initial lunar touchdown will be made by the lunar module crew to the CSM and to MSFN.

8-58. The crewman in the CSM will maintain visual observation of the lunar landing operation as long as possible. All three crewmen will be in their space suits.

8-59. <u>LUNAR SURFACE OPERATIONS</u>.

8-60. The lunar surface operations phase begins with lunar touchdown and ends with launching of the lunar module from the moon.

8-61. Initial tasks to be performed by the two astronauts following touchdown include review and determination of the lunar ascent sequence and of the parameters required. A complete check of the lunar module systems and structure will be made. Necessary maintenance will be determined and performed to assure the operational ascent capability of the lunar module. The systems will be put into a lunar-stay mode and a systems monitoring procedure established. The lunar module will be effectively secured as necessary, and the landing and launch stage disconnect mechanisms activated.

8-62. The lunar landing must be made on the earth-side of the moon to permit and establish communication with MSFN and the lunar orbiting CSM from the surface of the moon. Voice and signal communication will be verified prior to beginning egress and lunar exploration activity. A post-touchdown status report will be made to the CSM before line-of-sight communication is lost as the spacecraft orbits below the lunar horizon. The position and attitude of the lunar module on the moon will be established and reported.

8-63. The lunar module is capable of operating normally on the lunar surface during any phase of the lunar day-night cycle. The lunar module, designed to be left unoccupied with the cabin unpressurized on the lunar surface, will be capable of performing its operations independently of earth-based information or control.

8-64. Although the nominal lunar stay-time may be from 4 to 35 hours, depending on the planned scientific exploration program, the capability to launch at any time in an emergency situation shall be established. Portable life support systems will provide the capability for 24 man-hours of separation from the lunar module. Maximum continuous separation will be 4 hours (3 hours of normal operation plus 1 hour for contingencies). The astronauts shall alternately share the lunar surface exploration within the above mentioned PLSS limits until the exploration is concluded.

A0073   A0074   SM-2A-521B

Figure 8-15. Lunar Surface Operations

8-65. One of the two astronauts will descend to the lunar surface to perform scientific exploration and observation and will stay within sight of the crewmember remaining behind. The scientific exploration activity may include gathering selected samples from the lunar surface and atmosphere, measurement of lunar surface and atmospheric phenomena, and the securing of scientific instruments on the lunar surface for signal transmission and telescopic observation from earth. Video transmission from the lunar surface may be accomplished by means of portable television equipment. Provision will be made for return of approximately 80 pounds of samples from the lunar surface.

8-66. Following completion of the lunar surface exploration activity, preparation for ascent will begin. The two astronauts will return to the LM and secure themselves in the LM cabin. Launch and rendezvous plans will be confirmed with the crewman in the CSM and with MSFN. The spacecraft tracking and rendezvous data determination sequence will be initiated. The lunar module operational systems required for lunar ascent, the ascent and descent stage separation, and the inertial measurement unit alignment will be checked out.

8-67. CSM SOLO LUNAR ORBIT OPERATIONS.

8-68. During separation of the lunar module from the CSM for lunar operations, the crewmember in the CSM will perform a series of backup operations in support of the lunar activity.

8-69. The CSM crewmember will initially monitor the separation sequence and initiate optical tracking of the LM. A communication link between the CSM, lunar module, and MSFN will be established. The spacecraft will monitor the LM orbit injection sequence and maintain cognizance of essential operational parameters.

8-70. In addition, periodic operational checks will be made of the CSM systems, the inertial measurement unit alignment procedure performed, and the lunar orbit parameters periodically updated and confirmed with MSFN.

8-71. The lunar landing sequence may be monitored by optical tracking and essential operational data will be transmitted to MSFN. The location of the lunar landing site will be determined. Visual observations will be made of the lunar surface operations and periodic line-of-site communication with the crew maintained as required.

8-72. CSM confirmation of the LM ascent and rendezvous parameters will be established. Radar tracking of the ascent trajectory will be established to permit determination of the operational maneuvers to effect rendezvous. The spacecraft guidance and navigation system determines these essential parameters.

8-73. For rendezvous and docking, the CSM normally will be stabilized in a passive mode. However, the CSM will have the capability of controlling the terminal attitude and translation required for rendezvous and docking. The CSM solo operation ends after completion of docking, when the crew transfers from the LM to the CSM.

8-74. <u>LUNAR ASCENT.</u>

8-75. Confirmation of "go" conditions for lunar launch is made with the CSM and MSFN. The ascent engine will be ignited and launch from the lunar surface accomplished.

8-76. The lunar module ascent trajectory places it in a position approximately 50,000 feet above the lunar surface, at a velocity such that the resultant orbit about the moon has a clear minimum altitude of approximately 50,000 feet and a nominal intercept with the orbiting CSM. The launch trajectory of the lunar module is controlled through its guidance, navigation, and control system. The lunar module rendezvous radar tracks the CSM during the ascent to provide inputs to the guidance, navigation, and control system. The lunar module reaction control system executes the required roll, attitude, and pitch maneuvers to place it in the required orbit.

8-77. Following the cutoff of the ascent engine, the radar will continue to track the CSM. The CSM guidance and navigation system will compute the orbit of the lunar module and the crew will determine if any ΔV corrections are required to effect rendezvous. The final coast trajectory parameters, range, rate, and attitude angles will be determined and rendezvous operations will be initiated.

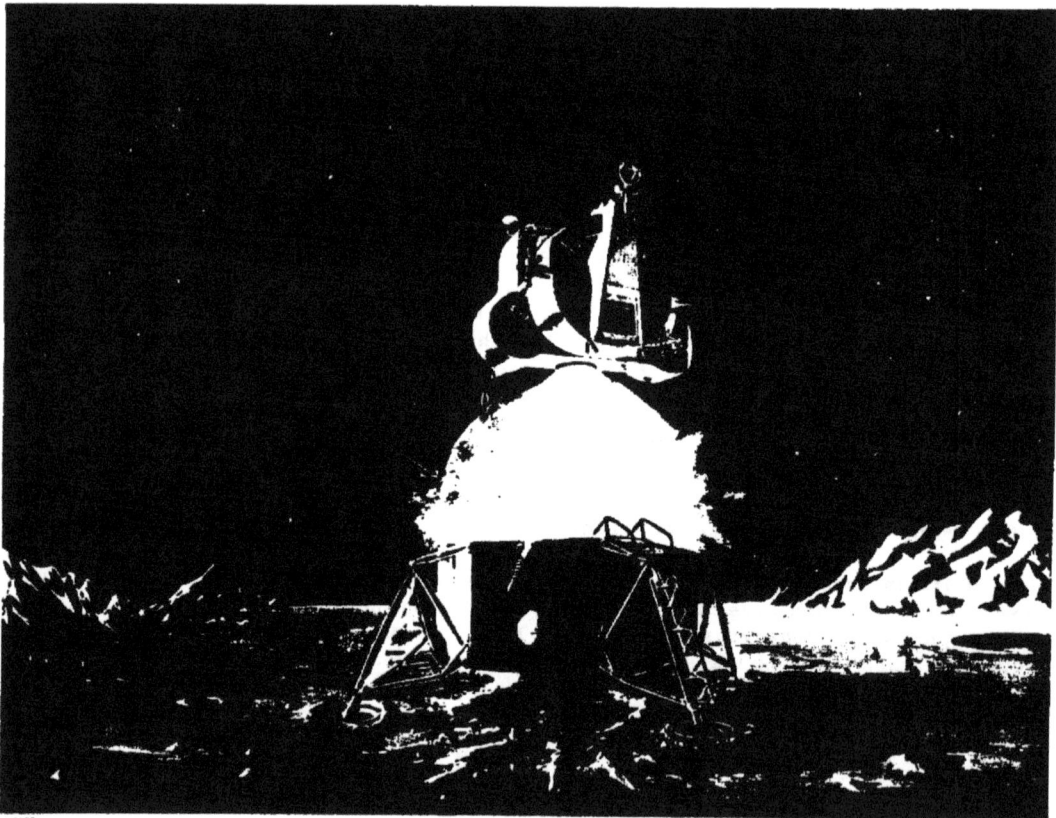

A0103

SM-2A-522C

Figure 8-16. Lunar Ascent

Figure 8-17. Rendezvous

8-78. RENDEZVOUS.

8-79. The rendezvous operations begin during the LM ascent phase. The ascent trajectory may require up to three midcourse corrections to reach a target course with the CSM. These corrections will be made with either the LM ascent engine or the reaction control system. The final rendezvous maneuvers will include three terminal homing thrusts from the LM reaction control system to reduce the relative velocity to a minimum.

8-80. The LM crew will manually control the lunar module within a range of approximately 500 feet from the CSM, with a relative velocity of 5 feet per second, or less. Both the CSM and the LM are capable of performing the final rendezvous and docking maneuvers required. The CSM is normally stabilized in passive mode with the lunar module operationally active to effect rendezvous.

8-81. The final rendezvous maneuvers of the LM to effect contact with the CSM, are performed using the reaction control engines. Docking alignment and closing velocity will be verified and necessary manual operational control established to effect engagement of the drogue and probe. Following verification of the drogue and probe engagement the latching sequence will be performed. Completion of final docking will be verified by both the LM and CSM crewmembers. The postdocking status of the system will be determined and, when the docking maneuver is completed, information will be transmitted to manned space-flight network.

8-82. Following engagement of the drogue and probe, and four semiautomatic latches (initial docking), an LM crewmember will remove the LM upper hatch and secure the LM to the CSM by locking eight manual latches and four semiautomatic latches provided (final docking). (See figure 3-22.) After the drogue and probe are removed and stowed in the LM, the pressures are allowed to equalize and the C/M access hatch is removed.

8-83. Following completion of final docking, the LM systems will be secured. The LM crew will then transfer the lunar scientific equipment and samples to the command module and store them. The two crewmembers will enter the command module and secure the access hatch between the command module and the lunar module.

8-84. A system status check will be made of the spacecraft operational systems and the status for separation from the LM will be communicated to MSFN. The sequence for separation will then be initiated. The LM will be pyrotechnically released from the CSM and the service module reaction control engines activated to translate the CSM away from the lunar module.

8-85. Following this separation, the CSM will be operationally maneuvered for the inertial measurement unit alignment. After fine-alignment of the inertial measurement unit, a series of three navigational sightings will be made to establish the transearth injection parameters. The transearth injection trajectory is computed by the Apollo guidance computer and subsequently confirmed with MSFN. The final transearth injection operational parameters, service module reaction control system ullage acceleration, required service propulsion system firing time, incremental velocity, thrust vector, and transearth injection attitude are determined and confirmed by MSFN.

8-86. TRANSEARTH INJECTION AND COAST.

8-87. The transearth injection phase covers the period of time the service propulsion system burns when injecting the CSM into a transearth trajectory.

8-88. For each lunar orbit there exists one opportunity for transearth injection. Injection occurs behind the moon, with respect to the earth, nominally one orbit after completion of rendezvous with the lunar module. The service propulsion system ullage acceleration is manually initiated to begin the earth injection sequence. The injection velocity increment for the predetermined transit time to earth is initiated with the service propulsion system thrust controlled by the guidance and navigation system.

8-89. The transearth coast phase begins with service propulsion system engine cutoff following transearth injection and ends at the entry interface altitude of 400,000 feet.

8-90. The transearth injection will be operationally performed to place the spacecraft in a return trajectory toward the earth, and will require a minimum of operational maneuvers and corrections. A ΔV budget sufficient to provide a total velocity correction of 300 feet per second during the transearth coast phase is provided. Nominally, three transearth velocity corrections may be made: one near the moon, the second approximately near the midpoint of the return trajectory, and the third ΔV increment near the earth.

9-91. The primary operations which occur consist of periodic systems checks, trajectory verification, determination of the ΔV corrections required, preparation for jettison of the service module, service module jettison from the command module, and preparations for earth entry.

8-92. The midcourse corrections are determined by means of sequential trajectory verifications. The MSFN computes variations from the required trajectory parameters; and determines velocity changes (if necessary), thrust vectors, and firing time. This data is confirmed with the Apollo guidance computer. The ΔV is operationally implemented and the verification of the velocity correction is subsequently determined after each midcourse correction.

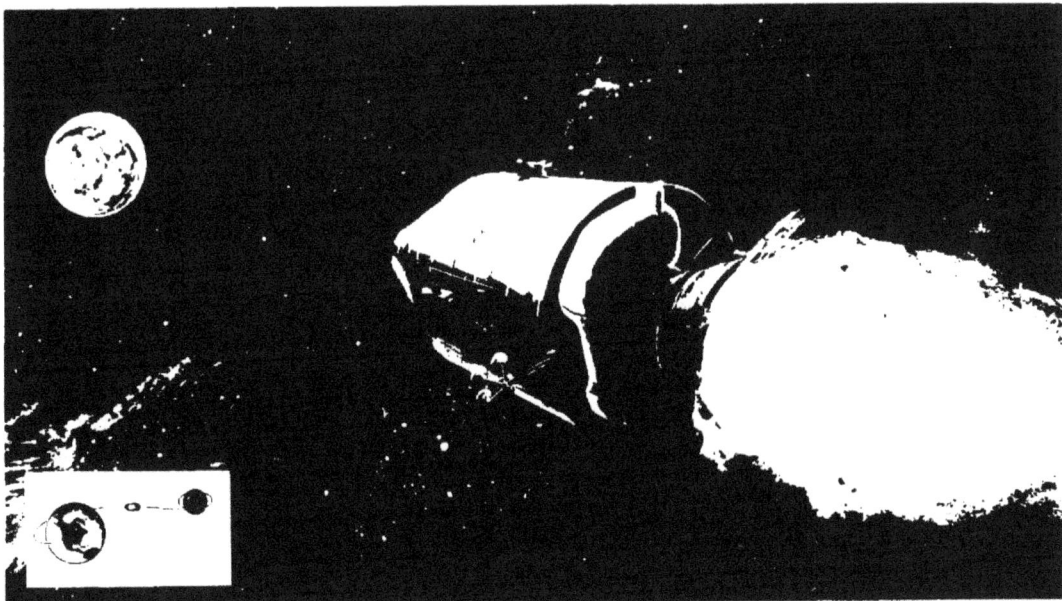

A0079   SM-2A-524B

Figure 8-18. Transearth Injection and Coast

SM-2A-525C

Figure 8-19. Service Module Jettison

8-93. SERVICE MODULE JETTISON.

8-94. Following the last midcourse correction, preparatory activity will be initiated for jettisoning the service module. The near-earth entry corridor and pre-entry parameters for separation of the service module from the command module will be determined by MSFN and confirmed by the Apollo guidance computer. The final systems check will be performed and the spacecraft oriented for service module jettison. The command module entry batteries will be activated, and the service module operationally jettisoned by pyrotechnic separation of the adapter and subsequent translational thrust by the service module reaction control engines to effect space separation of the command module and service module. The command module is then oriented into an MSFN-confirmed entry attitude by the command module reaction control engines.

8-95. A status check will be made of the systems after service module separation. Final operational checks will be made of the systems for entry. Confirmation of the entry parameters will be made with MSFN. The entry monitor control and display will be operationally activated, entry alignment of the inertial measurement unit made, and utilization of the flight director attitude indicator and Apollo guidance computer implemented.

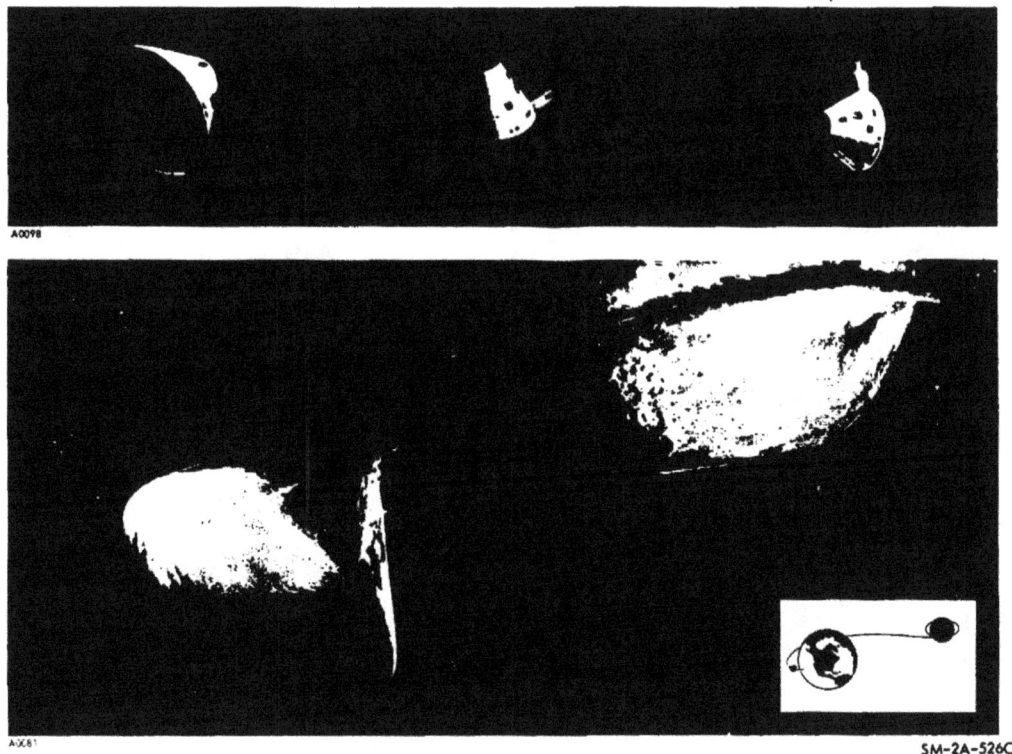

Figure 8-20. Earth Entry

8-96. <u>EARTH ENTRY.</u>

8-97. The earth entry phase begins at an altitude of 400,000 feet and ends upon activation of the earth landing system.

8-98. The operational control of the entry is dependent on the range required from the 400,000-foot entry point to the landing area. For short-entry ranges, no skip-out maneuver is required. For entry ranges approaching the upper limit, a skip-out maneuver is required to attain the greater distance. In either case, the lift maneuver is controlled by rolling the command module using the reaction control engines. Operational control is normally maintained through the guidance and navigation system with the commander providing a backup capability using the entry monitor display.

8-99. Entry into the earth atmosphere is sensed by a 0.05G signal indication. The entry attitude is determined on the flight director attitude indicator, and the entry monitor control display is observed. The Apollo guidance computer computes the range to "go" and provides navigation from the 0.05G point and time.

8-100. The range control maneuver is initiated by reaction control engines roll control and necessary pitch and yaw damping. The entry monitor display indicates the ΔV and G-level, and the survival display requirements. The guidance and navigation system executes the required reaction control system roll commands.

Figure 8-21. Earth Landing

8-101. EARTH LANDING.

8-102. The earth landing phase begins when the earth landing system is operationally armed at an altitude of 100,000 feet and ends with touchdown. The earth landing system automatic sequencer pyrotechnically ejects the forward heat shield at approximately 24,000 feet. Two seconds later, the two drogue parachutes are mortar-deployed in a reefed condition. The reefing lines are pyrotechnically severed and the drogue parachutes open fully in approximately 8 seconds to orient the C/M apex upward during descent to 10,000 feet. Three pilot parachutes are automatically mortar-deployed at 10,000 feet, and the drogue parachutes are pyrotechnically disconnected. The pilot parachutes in turn, deploy the three main parachutes to a line-stretch, reefed condition. The reefing lines are pyrotechnically severed and the main parachutes open fully in approximately 8 seconds. The main parachutes lower the command module to touchdown and impact at a terminal descent velocity, assuring an impact G-level consistent with the safety of the crew. The main parachute attach lines are pyrotechnically severed upon touchdown.

8-103. During the final part of the main parachute descent, the recovery communication systems are activated and transmit a location signal for reception by the operational recovery forces.

Figure 8-22. Recovery Operations, Primary Landing

8-104. RECOVERY OPERATIONS.

8-105. The recovery operations phase begins with touchdown and ends with the recovery of the crew and retrieval of the command module. The HF communication system is deployed and begins transmitting a repeating location signal for reception by the recovery task forces deployed in the area of predicted touchdown. Voice communication capability is also provided by the H-F communications system.

8-106. If touchdown occurs in water (primary landing), fluorescent dye goes into solution, coloring the water a bright, fluorescent, yellow-green over an extended area and lasting approximately 12 hours. The dye should be visible to recovery force aircraft or ships for a considerable distance. A flashing beacon light is also provided for use at night.

8-107. Immediately following a water landing, after the main parachutes are pyro-technically cut from the command module, the crew will assess the flotation status and capability of the command module. If the command module is in the inverted (stable II) flotation attitude, the crew will activate the uprighting system. When the command module achieves an upright (stable I) flotation attitude, the crew will remain in the command module or, if necessary, will leave in the inflatable liferaft provided for the three crew-members. Steps will be taken, as necessary, to effectively secure the command module for optimum flotation stability and subsequent retrieval. The capability is to be provided for helicopter pickup of the command module, using the recovery pickup loop, or a nearby ship may pick up the command module. The three crewmembers may be picked up by helicopter, ship, or boat. Land ground forces may pick up the command module if the land touchdown point is in an accessible area.

8-108. The flotation design will provide a survivable flotation capability for a minimum of 48 hours, under design sea conditions. A water landing provides fewer touchdown hazards and a correspondingly greater safety for the crew.

SM-2A-528

Figure 8-23. Recovery Operations, Backup Landing

APOLLO SUPPORT MANUALS

A-1. GENERAL.

A-2. Apollo support manuals consist of published data packages to support the Apollo program. The manuals are categorized into general series and defined by specific letter/number combinations as follows:

- SM1A-1   Index of Apollo Support Manuals and Procedures

- SM2A-02   Apollo Spacecraft Familiarization Manual

- SM2A-03-(S/C No.)   Preliminary Apollo Operations Handbook, Command and Service Module

- SM2A-03A-(S/C No.)   Preliminary Apollo Operations Handbook, Command and Service Module (Confidential Supplement to SM2A-03)

- SM2A-08-(S/C No.)   Apollo Recovery and Postlanding Operations Handbook

- SM3A-200   Apollo Ground Support Equipment Catalog

- SM6A-(Series No.)   Apollo Training Equipment Maintenance Handbooks

    -22   Electrical Power System Trainer

    -23   Environmental Control System Trainer

    -24   Stabilization Control System Trainer

    -25   Sequential Flow System Trainer

    -26   Propulsion System Trainer

    -41-1 and -2   Mission Simulators Maintenance and Operations Manual

- SM6T-2-02   Apollo Mission Simulator Instructor Handbook

A-3. INDEX OF APOLLO SUPPORT MANUALS AND PROCEDURES.

A-4. Index SM1A-1 is published periodically and provides a listing of all Apollo Support Manuals and Procedures in publication.

A-5. APOLLO SPACECRAFT FAMILIARIZATION MANUAL.

A-6. The familiarization manual (SM2A-02) presents a general, overall description of the Apollo program. Coverage includes physical configuration, functional operation, the test program, and the missions of the equipment utilized within the scope of the Apollo program. General terms are used in the descriptive text with sufficient detail to ensure comprehension.

A-7. PRELIMINARY APOLLO OPERATIONS HANDBOOK, COMMAND AND SERVICE MODULE.

A-8. A preliminary Apollo operations handbook (SM2A-03 and -03A) is a preliminary version of the Apollo Operations Handbook designed as a single source of spacecraft data and operational procedures. The handbook provides detailed spacecraft operating instructions and procedures for use by the crew during all phases of the latest, manned Block I mission. This involves normal, alternate, backup, malfunction, and contingency procedures, crew procedures, and checklists. CSM and LM interface instructions and procedures applicable to the command module are included, with information on spacecraft controls and displays, systems data, crew personal equipment, in-flight experiments, and scientific equipment.

A-9. APOLLO RECOVERY AND POSTLANDING OPERATIONS HANDBOOK.

A-10. The Apollo recovery and postlanding operations handbook (SM2A-08-S/C No.) provides detailed information for retrieval of the recoverable portion of the spacecraft. Descriptive text and illustrations specify procedures and provide the information necessary to perform recovery and postlanding operations of each manned and unmanned command module scheduled for recovery.

A-11. APOLLO GROUND SUPPORT EQUIPMENT CATALOG.

A-12. An Apollo ground support equipment catalog (SM3A-200) is provided to identify and illustrate items of specific auxiliary, checkout, handling, and servicing equipment associated with Project Apollo.

A-13. APOLLO TRAINING EQUIPMENT MAINTENANCE HANDBOOKS.

A-14. An Apollo training equipment maintenance handbook (SM6A-series) is provided for the maintenance of the systems trainers and mission simulators.

A-15. APOLLO MISSION SIMULATOR INSTRUCTOR HANDBOOK.

A-16. An instructor's handbook (SM6T2-02) is provided for the mission simulators. The handbook provides the necessary information for training astronauts on the Apollo mission simulators.

GLOSSARY OF ABBREVIATIONS, SYMBOLS, AND TERMS

This glossary lists terminology found in Apollo documentation and engineering drawings. Frequently used common terms, which are industry standard, have been omitted for brevity. This glossary will be updated to reflect the latest changes during each revision of the manual.

ABBREVIATIONS

AAO	Astronauts Activities Office	AERO-A	AERO - Aerodynamics analysis
ABD	Airborne Ballistics Division (NASA)	AERO-D	AERO - Dynamics analysis
AC	Audio center	AERO-DIR	AERO - Director
ACA	Associate contractor administration	AERO-E	AERO - Experimental aerodynamics
ACE	Acceptance checkout equipment	AERO-F	AERO - Flight evaluation
ACED	AC Electronics Division	AERO-G	AERO - Aerophysics and astrophysics
ACF	American Car and Foundry		
ACM	Audio center module	AERO-P	AERO - Future projects
ACME	Attitude control and maneuvering electronics	AERO-PCA	AERO - Program coordination and administration
A&CO	Assembly and checkout		
ACR	Associate contractor	AERO-PS	AERO - Projects staff
ACRC	Audio center - receiver	AERO-TS	AERO - Technical and scientific staff
ACS	Attitude control and stabilization	AF	Audio frequency
ACSB	Apollo crew systems branch	AFCS	Automatic flight control system
ACTM	Audio center - transmitter		
ACV	AC volts	AFETR	Air Force eastern test range
AD	Apollo development	AGAA	Attitude gyro accelerometer assembly
A/D	Analog-to-digital		
ADA	Angular differentiating accelerometer	AGANI	Apollo guidance and navigation information
ADC	Analog-to-digital converter	AGAP	Attitude gyro accelerometer package (superseded by AGAA)
ADF	Automatic direction finding (equip.)		
ADP	Automatic data processing	AGC	Apollo Guidance Computer (used by MIT)
AEB	Aft equipment bay		
AEDC	Arnold Engineering Development Center	AGC	Aerojet General Corporation
		AGCS	Automatic Ground Control Station (NASA)
AERO	Aeroballistics (MSFC)		
		AGCU	Attitude gyro coupling unit

AGE	Apollo Guidance and Navigation Equipment (used by MIT)	APHFFF	Ames Prototype Hypersonic Free Flight Facility
AGE	Aerospace ground equipment	APK	Accelerometer package
AIAA	American Institute of Aeronautics & Astronautics	APO	Apollo project office
		APP	Access point PACE
AIDE	Aerospace installation diagnostic equipment	APTT	Apollo Part Task Trainer
		APU	Auxiliary power unit
AII	Apollo implementing instructions (NAA, S&ID)	AQ	Apollo qualification
		ARA	Auxiliary recovery antenna
ALFA	Air lubricated free attitude	ARC	Ames Research Center (NASA) (Moffet Field, Calif.)
ALIAS	Algebraic logic investigations of Apollo systems		
		ARE	Apollo reliability engineering
AM	Amplitude modulation	AREE	Apollo reliability engineering electronics
AMG	Angle of middle gimbal		
AMMP	Apollo master measurements program	ARIS	Advanced range instrumentation ships
AMOO	Aerospace Medical Operations Office (MSC)	ARM	Apollo Requirements Manual
		ARS	Attitude reference system
AMPTF	Apollo Mission Planning Task Force	ASCS	Automatic stabilization and control system
AMR	Atlantic Missile Range (Superseded by ETR)	ASDD	Apollo signal definition document
AMRO	Atlantic Missile Range Operations (MSC)	ASDTP	Apollo Spacecraft Development Test Plan
AMS	Acoustic measurement system	ASFTS	Auxiliary systems function test stand
AMS	Apollo mission simulator		
AMW	Angular momentum wheel	AS/GPD	Attitude set and gimbal position display
AOH-CSM	Apollo Operations Handbook-CSM		
		ASI	Apollo systems integration
AOH-LM	Apollo Operations Handbook-LM	ASM	Apollo Systems Manual
		ASP	Apollo spacecraft project
AORA	Atlantic Ocean Recovery Area	ASPI	Apollo supplemental procedural information
AOS	Atlantic Ocean ship		
AP	Access point	ASPO	Apollo Spacecraft Project Office
AP	Accelerometer package		
APC	Procurement and Contracts Division (MSC)	ASTR	Astronics (MSFC)
		ASTR-A	ASTR - Advanced studies
APCA	Procurement Apollo	ASTR-ADM	ASTR - Administrative
APCAL	Procurement Apollo lunar module	ASTR-E	ASTR - Electrical systems integration
APCAN	Procurement Apollo navigation and guidance	ASTR-F	ASTR - Flight dynamics
		ASTR-G	ASTR - Gyro and stabilizer
APCAS	Procurement Apollo spacecraft	ASTR-I	ASTR - Instrumentation development
APCAT	Procurement Apollo test and instrumentation		
		ASTR-M	ASTR - Electromechanical engineering
APCR	Apollo program control room		
APD	Advanced program development	ASTR-N	ASTR - Guidance and control systems

ASTR-P	ASTR - Pilot manufacturing development	CBC	Complete blood count
ASTR-PC	ASTR - Program coordination	CBX	C-band transponder
ASTR-R	ASTR - Applied research	cc	Cubic centimeter
ASTR-TSA	ASTR - Advanced research and technology	CCTV	Closed-circuit television
		CCW	Counterclockwise
ASTR-TSJ	ASTR - Saturn	C/D	Countdown
ASTR-TSR	ASTR - Reliability	C&D	Communication and data
ATC	Assistant test conductor	CDC	Computer Development Center (NASA)
ATO	Apollo Test and Operations	CDCM	Coupling display manual control - IMU
AT&O	Apollo Test and Operations (ATO is preferred)	CDCO	Coupling display manual control - optics
ATR	Apollo test requirement	CDOH	Coupling display optical hand controller
ATS	Atlantic tracking ship		
AUTO	Automatic	CDR	Critical design review
A-V	Audio-visual	CDRD	Computations and Data Reduction Division (MSC)
AVC	Automatic volume control		
AVSS	Apollo Vehicle Systems Section (NASA)	CDSC	Coupling display SCT manual control
AWI	Accommodation weight investigation	C&DSS	Communications and data subsystems
		CDU	Coupling display unit
BAC	Boeing Aircraft Company	CDU	Coupling data unit
BATT	Battery	CDUM	Coupling display unit - IMU
BCD	Binary coded decimal	CDUO	Coupling display unit - optics
BCO	Booster engine cutoff	CEPS	Command module electrical power system
BDA	Bermuda (remote site)		
BECO	Booster engine cutoff	C/F	Center frequency
BER	Bit error rate	CFAE	Contractor-furnished airborne equipment
BG	Background		
BLWR	Blower	CFD	Cumulative frequency distribution
BM	Bench maintenance		
B/M	Bench maintenance (BM is preferred)	CFE	Contractor-furnished equipment
BMAG	Body-mounted attitude gyro	CFDF	Crew flight data file
BME	Bench maintenance equipment	CFM	Cubic feet per minute
BMG	Body-mounted (attitude) gyro (BMAG is preferred)	cg	Center of gravity
		CGSS	Cryogenic gas storage system
BOA	Broad ocean area		
BOD	Beneficial occupancy data	CH4	Methane
BP	Blood pressure	CHGE	Charger
BP	Boilerplate	C&I	Communication and Instrumentation
BPC	Boost protective cover		
BPS	Bits per second	CIF	Central Information Facility (AMR)
BSI	Booster situation indicator		
B/U	Backup	CIR&SEP	H2 Circulation, water separation centrifuge, and glycol circulation
COAS	Crewman optical alignment sight		
		CIS	Communication and instrumentation system
C/B	Circuit breaker		
CBA	C-band transponder antenna		

C&IS	Communication and instrumentation system (CIS is preferred)	DBM	Decibels with respect to one milliwatt
CL	Closed-loop	DBW	Decibels with respect to one watt
CLM	Circumlunar mission	D&C	Displays and controls
C/M	Command module	DCA	Design change authorization
CMM	Communications and Telemetry (used by MIT)	DCCU	Decommutator conditioning unit (PACE)
CO	Carbon monoxide	DCIB	Data communication input buffer
C/O	Checkout	DCOS	Data communication output selector
C/O	Cutoff		
CO2	Carbon dioxide	DCS	Design control specification
COMP	Compressor	DCU	Display and control unit
CP	Control panel	DCV	DC volts
CP	Control Programer	DDP	Data distribution panel
CPE	Chief project engineer	DDS	Data display system
CPEO	CPE Engineering Order	DE	Display electronics
CPO	Central Planning Office (MSFC)	DEA	Display electronics assemblies
cps	Cycles per second	DECA	Display/AGAP electronic control assembly
CPS	Critical path schedule		
CRT	Cathode-ray tube	DEI	Design engineering inspection
CRYO	Cryogenics	DF	Direction finding
CS	Communication system	D/F	Direction finder
CSD	Computer systems director	DFS	Dynamic flight simulator
CSD	Crew System Division (MSC)	DIM	Design information manual
CSM	Command and service module	DISC	Discharge
CSS	Crew safety system	DISPLAY/ AGAA ECA	Display and attitude gyro, accelerometer assembly-electronic control assembly
CSS	Cryogenic storage system		
CST	Combined systems test		
CSTU	Combined systems test unit	DM	Design manual
CTE	Central timing equipment	DNR	Downrange
CTL	Component Test Laboratory (NASA)	DOD	Department of Defense
		DOF	Degree of freedom
CTN	Canton Island (remote site)	DOF	Direction of flight
CTU	Central timing unit	DOVAP	Doppler velocity and position
CUE	Command uplink electronics (ACE)	DP	Design proof
		DPC	Data processing center
CW	Clockwise	DPDT	Double-pole double-throw
CW	Continuous wave	DPST	Double-pole single-throw
CWG	Constant-wear garment	DRM	Drawing requirements manual
CYI	Canary Islands	DSB	Double sideband
		DSE	Data storage equipment
DA	Dip angle	DSIF	Deep-space instrumentation facility
DA	Double amplitude		
DAC	Digital-to-analog converter	DSKY	Display and keyboard
DAE	Data acquisition equipment	DTCS	Digital test command system (ACE)
DART	Director and response tester		
DAS	Data acquisition system		
db	Decibel		

DTS	Data transmission system	ELS	Earth landing system
DTVC	Digital transmission and verification converter (ACE)	E/M	Escape motor
		EMD	Entry monitor display
DVD	Delta velocity display	EMG	Electromyograph, electromyography, electromyogram
DVO	Delta velocity on/off		
DVU	Delta velocity ullage	EMI	Electromagnetic interference
		EMS	Entry monitor system
EBW	Explosive bridgewire	ENVR	Environmental
ECA	Electronic control assembly	EO	Earth orbit
ECA	Engineering change analysis	EO	Engineering order
ECAR	Electronic control assembly - roll	E&O	Engineering and operations (building)
ECD	Engineering control drawing	EOD	Explosive ordnance disposal
ECD	Entry corridor display	EOL	Earth orbit launch
ECET	Electronic control assembly - engine thrust	EOM	Earth orbital mission
		EOR	Earth orbital rendezvous
ECG	Electrocardiograph, electrocardiography, electrocardiogram	EPDS	Electrical power distribution system
		EPS	Electrical power system
ECK	Emergency communications key	EPSTF	Electrical power system test facility
ECN	Engineering change notice		
ECO	Engine cutoff	EPUT	Events per unit time
ECO	Engineering Change Order (MSC)	ERG	Electroretinograph, electroretinography, electroretinogram
ECPY	Electronic control assembly - pitch & yaw		
		ERP	Eye reference point
ECS	Environmental control system	ERS	Earth recovery system
ECU	Environmental control unit	ERU	Earth rate unit (15 degrees/ hour)
EDL	Engineering development laboratories (NAA, S&ID)		
		ESB	Electrical Systems Branch (MSC)
EDP	Electronic data processing		
EDPM	Electronic data processing machine	ESE	Engineering support equipment
EDS	Emergency detection system	ESS	Emergency Survival System (NASA)
EED	Electroexplosive device		
EEG	Electroencephalograph, electroencephalography, electroencephalogram	ESS	Entry survival system
		ESTF	Electronic System Test Facility (NASA)
EET	Equivalent exposure time	ESV	Emergency shutoff valve
EFSSS	Engine failure sensing and shutdown system	ET	Escape tower
		E/T	Escape tower
EHF	Extremely high frequency	ETF	Eglin Test Facility
EI	Electromagnetic interference	ETOC	Estimated time of correction
EI	Electronic interface	ETR	Eastern test range
EKG	Electrocardiograph, electrocardiography, electrocardiogram	EVAP	Evaporator
		EVT	Extravehicular transfer
ELCA	Earth landing control area	FACT	Flight acceptance composite test

FAE	Final approach equipment	G&CEP	Guidance and control equipment performance
FAP	Fortran assembly program		
FAX	Facsimile transmission	GCU	(Attitude) gyro coupling unit (AGCU is preferred)
FC	Ferrite core		
F/C	Fuel cell	GDC	Gyro display coupler
F/C	Flight control	GETS	Ground equipment test set
FCD	Flight control division	GFAE	Government-furnished aeronautical equipment
FCH	Flight controller's handbook		
FCOB	Flight Crew Operations Branch (NASA)	GFE	Government-furnished equipment
FCOD	Flight Crew Operations Division (MSC)	GFP	Government-furnished property
FCSD	Flight Crew Support Division (MSC)	GG	Gas generator
		GH_2	Gaseous hydrogen
FCT	Flight Crew Trainer (IMCC)	GHE	Gaseous helium
FD	Flight Director (NASA)	GHe	Gaseous helium (preferred)
FDAI	Flight director attitude indicator	GLY	Glycol
		GMT	Greenwich Mean Time
FDO	Flight dynamics officer	G&N	Guidance and navigation
FDRI	Flight director rate indicator	G&NS	Guidance and navigation system
FEB	Forward equipment bay	GN_2	Gaseous nitrogen
FEO	Field engineering order	GNC	Guidance and navigation computer
FF	Florida Facility		
FHS	Forward heat shield	GNE	Guidance and navigation electronics
FIDO	Flight dynamics officer		
FLSC	Flexible linear-shaped charge	GO_2	Gaseous oxygen
FM	Frequency modulation	GORP	Ground operational requirements plan
FMA	Failure mode analysis		
FMD&C	Flight mechanics, dynamics and control	GOSS	Ground Operational Support System (superseded by MSFN)
FMX	FM transmitter	GOX	Gaseous oxygen (superseded by GO_2)
FO	Florida Operations		
FOD	Flight Operations Division (MSC)	GP	General purpose
		GPD	Gimbal position display
FOF	Flight operations facilities	GPI	Gimbal position indicator
FORTRAN	Formula translation	gpm	Gallons per minute
FOS	Flight operations support	GSDS	Goldstone duplicate standard (standard DSIF equipment)
FP	Fuel pressure		
FPO	Future Projects Office (MSFC)	GSE	Ground support equipment
		GSFC	Goddard Space Flight Center (NASA) (Greenbelt, Md.)
FPS	Frames per second		
fps	Feet per second	GSP	Guidance signal processor
FQ	Flight qualification	GSPO	Ground Systems Project Office (MSC)
FQR	Flight qualification recorder		
FRDI	Flight research and development instrumentation	GSP-R	Guidance signal processor - repeater
		GSR	Galvanic skin response
FRF	Flight readiness firing	GSSC	General Systems Simulation Center (NASA)
FSK	Frequency shift keying		
FTP	Flight test procedure	GSSC	Ground Support Simulation Computer (MCC)
GAEC	Grumman Aircraft Engineering Corp.		
		GTI	Grand Turk Island
GC	Gigacycles (1000 megacycles)	GTK	Grand Turk
G&C	Guidance and control		

GTP	General test plan	I/C	Intercom
G vs T	Deceleration units of gravity versus time	ICA	Item change analysis
		ICD	Interface control document
G vs V	Deceleration units of gravity versus velocity	ICM	Instrumentation and Communications monitor
GYI	Grand Canary Island (remote site)	ID	Inside diameter
		IESD	Instrumentation and Electronic Systems Division (MSC)
GYM	Guaymas, Mexico (remote site)	IF	Intermediate frequency
		I/F	Interface
H_2	Hydrogen	IFM	In-flight maintenance
H/A	Hazardous area	IFT	In-flight test
HAA	High altitude abort	IFTM	In-flight test and maintenance
HAW	Kauai Island, Hawaii (remote site)	IFTS	In-flight test system
		IG	Inner gimbal
HBW	Hot bridgewire	IGA	Inner gimbal axis
HC	Hand control	IL	Instrumentation Lab (MIT)
He	Helium	IL	Inertial Lab (NASA)
H/E	Heat exchanger	IL	Internal letter
HF	High frequency	ILCC	Integrated launch checkout and control
H/F	Human factors		
HFA	High frequency recovery antenna	IMCC	Integrated Mission Control Center (superseded by MCC)
HFX	High frequency transceiver	IMU	Inertial measurement unit
		IND	Indicator
HGA	High-gain antenna (2 KMC)	INS	Inertial navigation system
		INV	Inverter
HGB	Hemoglobin	IORA	Indian Ocean Recovery Area
HI	High	IOS	Indian Ocean ship (tracking)
H_2O	Water	I/P	Impact predictor
H_2S	Hydrogen sulfide	IR	Infra red
HS	Hot short	IRG	Inertial rate gyro
HS	Hydrogen sulfide	IRIG	Inertial rate integrating gyroscope (used by MIT)
H/S	Heat shield		
H-S	Hamilton Standard	IRP	Inertial reference package
H-S	Horizon scanner	IS	Instrumentation system
HS/C	House spacecraft	I_{sp}	Specific impulse
HSD	High-speed data	IST	Integrated systems test
HTRS	Heaters	IU	Instrumentation unit
HW	Hotwire	I/U	Instrumentation unit (IU is preferred)
HWT	Hypersonic wind tunnel		
		IUA	Inertial unit assembly
H/X	Heat exchanger		
		J/M	Jettison motor
IA	Input axis		
IAD	Interface analysis document	KC	Kilocycle (1000 cycles per second)
IAS	Indicated air speed		
IC	Intercommunication equipment	KMC	Kilomegacycle (gigacycle)
		KNO	Kano, Nigeria (remote site)
		KOH	Potassium hydroxide

KSC	Kennedy Space Center	LN$_2$	Liquid nitrogen
KW	Kilowatt	LO	Launch operations
		LO	Low
LAC	Lockheed Aircraft Corporation	L/O	Lift-off
LAET	Limiting actual exposure time	LO$_2$	Liquid oxygen
LC	Launch complex	LOC	Launch Operations Center
LC-39	Launch complex 39		(NASA)(Cocoa Beach, Fla.)
LCC	Launch Control Center (MCC)	LOD	Launch operations directorate
LCE	Launch complex engineer	LOD	Launch Operations Division
LCS	Launch control system		(NASA)(superseded by LOC)
L/D	Lift-drag ratio	LOM	Lunar orbital mission
LDGE	LM dummy guidance equipment	LOR	Lunar orbital rendezvous
		LOS	Line of sight
LDP	Local data package	LOS	Loss of signal
LDT	Level detector	LOX	Liquid oxygen (superseded by
LE	Launch escape		LO$_2$)
L/E	Launch escape (LE is preferred)	LP	Lower panel
		LPA	Log periodic antenna
LEB	Lower equipment bay	LPC	Lockheed Propulsion Company
LEC	Launch escape control	LPGE	LM partial guidance
LECA	Launch escape control area		equipment
LEM	Launch escape motor	LRC	Langley Research Center
LES	Launch escape system		(NASA)(Hampton, Va.)
LESC	Launch escape system control	LRC	Lewis Research Center
LET	Launch escape tower		(NASA)(Cleveland, Ohio)
LEV	Launch escape vehicle	LRD	Launch Recovery Division
LGC	LM guidance computer	LSB	Lower sideband
LGE	LM guidance equipment	LSC	Linear-shaped charge
LH	Left-hand	LSD	Low-speed data
LH$_2$	Liquid hydrogen	LSD	Life Systems Division (MSC)
LHA	Local hour angle		(superseded by CSD)
LHE	Liquid helium	LSD	Launch systems data
LHe	Liquid helium (preferred)	LSS	Life support system
LHFEB	Left-hand forward equipment bay	LTA	LM test article
		LTC	Launch vehicle test conductor
LHSC	Left-hand side console	LTDT	Langley transonic dynamics
LiOH	Lithium hydroxide		tunnel
LJ	Little Joe	LUPWT	Langley unitary plan wind
LL	Low-level		tunnel
LLM	Lunar landing mission	LUT	Launch Umbilical Tower
LLM	Lunar landing module	LV	Launch vehicle
LLOS	Landmark line of sight	LV	Local vertical
LLV	Lunar landing vehicle	L/V	Launch vehicle
LM	Landmark	LVO	Launch Vehicle Operations
LM	Lunar module		(MSFC)
L&M	Light and Medium Vehicles Office (MSFC)	LVOD	Launch Vehicle Operations Division (MSFC)
LMSC	Lockheed Missile and Space Company	LVSG	Launch vehicle study group

MAN	Manual	MN A	Main bus A
MASTIF	Multi-axis spin test inertial facility	MN B	Main bus B
M. C. & W. S.	Master caution and warning system	MNEE	Mission nonessential equipment
MCA	Main console assembly	M&O	Maintenance and operation
MCC	Mission control center	MOC	Master operations control
MCOP	Mission control operations panel	MOCR	Mission Operations Control Room (MCC)
MCR	Master change record	MODS	Manned Orbital Development Station (MSC)
MD	Master dimension	MORL	Manned orbiting research laboratory
MDC	Main display console	MOV	Main oxidizer valve
MDF	Main distribution frame (MCC)	MPTS	Multipurpose tool set
MDF	Mild detonating fuse	MRCR	Measurement requirement change request
MDR	Mission data reduction	MRO	Maintenance, repair, and operation
MDS	Malfunction detection system		
MDS	Master development schedule	M&S	Mapping and surveying
MDSS	Mission data support system	MSC	Manned Spacecraft Center (NASA) (Clear Lake, Texas)
MDT	Mean downtime		
MEC	Manual emergency controls	MSC-FO	Manned Spacecraft Center - Florida operations
MEE	Mission essential equipment	MSD	Mission systems data
MERu	Milli-earth rate unit (0.015 degree/hour)	MSFC	Marshall Space Flight Center (NASA) (Huntsville, Ala.)
MESC	Master event sequence controller	MSFC-LVO	Marshall Space Flight Center - Launch Vehicle Operations
MEV	Million electron volts	MSFN	Manned Space Flight Network (formerly GOSS)
MFG	Major functional group		
MFV	Main fuel valve	MSFP	Manned space flight program
MG	Middle gimbal	MT	Magnetic tape
MG	Motor-generator	MTF	Mississippi Test Facility (NASA)
MGA	Middle gimbal axis		
MGE	Maintenance ground equipment	MTS	Master timing system
MI	Minimum impulser	MTU	Magnetic tape unit
MIG	Metal inert gas	MTVC	Manual thrust vector control
MIL	Miliradian	MU	Mockup
MILA	Merritt Island Launch Area (superseded by KSC)	M/U	Mockup (MU is preferred)
MILPAS	Miscellaneous information listing program Apollo spacecraft	MUC	Muchea, Australia (remote site)
		MV	Millivolt
		MVD	Map and visual display (unit)
MIT	Massachusetts Institute of Technology	MW	Milliwatt
ML	Mold line	MWP	Maximum working pressure
MLT	Mission life test	N_2	Nitrogen
M/M	Maximum and minimum	NAA	North American Aviation
MMH	Monomethylhydrazine (fuel)	NAACD	NAA, Columbus Division
MMHg	Millimeters of mercury	NAARD	NAA, Rocketdyne Division
MMU	Midcourse measurement unit	NAASD	NAA, Space Division

NASA	National Aeronautics and Space Administration	ODM	One-day mission
N/B	Narrow band	O/F	Orbital flight
NC	Nose cone	O/F	Oxidizer-to-fuel ratio
N/C	Normally closed	OFB	Operational Facilities Branch (NASA)
N&G	Navigation and guidance (G&N is preferred)	OFO	Office of Flight Operations (NASA)
NH4	Ammonium	OG	Outer gimbal
N2H4	Hydrazine (fuel)	OGA	Outer gimbal axis
NM	Nautical mile	OIB	Operations Integration Branch (NASA)
NMO	Normal manual operation		
N/O	Normally open	OIS	Operational instrumentation system (MCC)
NOS ESS	Nonessential		
N2O4	Nitrogen tetroxide (oxidizer)	OL	Overload
NPC	NASA procurement circular	OL	Open-loop
NPDS	Nuclear particle detection system	OMSF	Office of Manned Space Flight (NASA)
NPSH	Net positive suction head	OMU	Optical measuring unit
NRD	National Range Division	OOA	Open ocean area
NRZ	Nonreturn to zero	OPS	Operations Director (NASA)
NSC	Navigational star catalog	OR	Operations requirements document (range user)
NSIF	Near Space Instrumentation Facility		
		OSS	Office of Space Sciences
NSM	Network status monitor	OTDA	Office of Tracking and Data Acquisition
NST	Network support team		
NTO	Nitrogen tetroxide (oxidizer)	OTP	Operational test procedure
NVB	Navigational base	OTU	Operating test unit
		OVERS	Orbital vehicle re-entry simulator
O2	Oxygen		
OA	Output axis	OXID	Oxidizer
OA	Ominiantenna		
OAM	Office of Aerospace Medicine (NASA)	ΔP	Pressure change (differential)
		PA	Precision angle (used by MIT)
OAO	Orbital astronomical observatory	PA	Power amplifier
		PA	Pad abort (used by NASA)
O&C	Operation and checkout	P/A	Pressure actuated
OCC	Operational control center	ACE	Automatic checkout equipment
OCDU	Optics coupling display unit (G&N)		
		PAFB	Patrick Air Force Base
O&C/O	Operation and checkout (O&C is preferred)	PAM	Pulse-amplitude modulation
		PATH	Performance analysis and test histories
OD	Operations directive		
OD	Outside diameter (overall diameter)	P&C	Procurement and Contracts (MSFC)
		P/C	Pitch control
ODDA	Office of Deputy Director for Administration (MSFC)	PCCP	Preliminary contract change proposal
ODDRD	Office of Deputy Director for Research and Development (MSFC)	PCD	Procurement control document
		PCM	Pitch control motor

PCM	Pulse-code modulation	PP	Partial pressure
PCME	Pulse-code modulation event	PPE	Premodulation processor equipment
PCPL	Proposed change point line		
PDA	Precision drive axis	PPM	Parts per million
PDD	Premodulation processor - deep-space data	PPS	Pulse per second
		PPS	Primary propulsion system
PDU	Pressure distribution unit	PR	Pulse rate
PDV	Premodulation processor - deep-space voice	PRA	Precession axis
		PRESS	Pressure
PE	Positive expulsion	PRF	Pulse repetition frequency
PE	Project engineer	PRM	Pulse-rate modulation
PEP	Peak envelope power	PRN	Pseudo-random noise
PERT	Program evaluation and review technique	PSA	Power and servo assembly
		PSA	Power servo amplifier
PF	Preflight	PSD	Phase sensitive demodulator
PFL	Propulsion Field Laboratory (Rocketdyne)	PSDF	Propulsion System Development Facility
PFM	Pulse-frequency modulation	psia	Pounds per square inch absolute
PFRT	Preliminary Flight Rating Test	psig	Pounds per square inch gage
PGA	Pressure garment assembly	PSK	Phase shift keyed
PGNCS	Primary guidance navigation control system	PSO	Pad safety officer (superseded by PSS)
pH	Alkalinity to acidity content (hydrogen ion concentration)	PSP	Program support plan
		PSS	Pad safety supervisor
PIAPACS	Psychophysical information acquisition processing and control system	PTPS	Propellant transfer pressurization system
		PTT	Push-to-talk
PIGA	Pendulous integrating gyroscopic accelerometer	PTV	Parachute test vehicle
		PU	Propellant utilization
PAO	Public Affairs Office NASA	PU	Propulsion Unit (NASA)
PIP	Pulsed integrating pendulous (accelerometer)	PUGS	Propellant utilization gauging system
PIPA	Pulsed integrating pendulous accelerometer	P&VE	Propulsion and Vehicle Engineering (MSFC)
PIRD	Project instrumentation requirement document	P&VE-ADM	P&VE - Administrative
		P&VE-DIR	P&VE - Director
PIV	Peak inverse voltage	P&VE-E	P&VE - Vehicle engineering
PL	Postlanding	P&VE-F	P&VE - Advanced flight systems
PLSS	Portable life support system		
PMP	Premodulation processor	P&VE-M	P&VE - Engineering materials
PMR	Pacific Missile Range	P&VE-N	P&VE - Nuclear vehicle projects
PND	Premodulation processor - near earth data	P&VE-O	P&VE - Engine management
POD	Preflight Operations Division (MSC)	P&VE-P	P&VE - Propulsion and mechanics
POI	Program of Instruction (NASA)	P&VE-PC	P&VE - Program coordination
POL	Petroleum oil and lubricants	P&VE-REL	P&VE - Reliability
POS	Pacific Ocean ship		

P&VE-S	P&VE - Structures	RGP	Rate gyro package
P&VE-TS	P&VE - Technical and		(superseded by RGA)
	scientific staff	RGS	Radio guidance system
P&VE-V	P&VE - Vehicle systems	RH	Relative humidity
	integration	RH	Right-hand
P&W	Pratt & Whitney	RHFEB	Right-hand forward equipment
P&WA	Pratt & Whitney Aircraft		bay
PYRO	Pyrotechnic	RHSC	Right-hand side console
		RJS	Reaction jet system
QA	Quality assurance	RO	Reliability Office (MSFC)
QAD	Quality Assurance Division	RP-1	Rocket propellant No. 1
	(MSFC)		(kerosene)
QAM	Quality assurance manual	RPD	Research Projects Division
QC	Quality control		(MSFC)
QD	Quick-disconnect	RR	Respiration rate
QRS	Qualification review sheet	RRS	Restraint release system
QUAL	Quality Assurance Division	RR/T	Rendezvous Radar/
	(MSFC)		Transponder
QVT	Quality verification testing	R/S	Range safety
		RSC	Range safety control
RA	Radar altimeter	RSC	Range safety command
RAD	Radiation absorbed dose	RSCIE	Remote station communication
RAE	Range, azimuth, and elevation		interface equipment
RAPO	Resident Apollo Project Office	RSO	Range safety officer
	(NASA)	RSS	Reactants supply system
RASPO	Resident Apollo Spacecraft	RTC	Real Time Computer (MCC)
	Project Office (MSC)	RTCC	Real Time Computer Complex
RB	Radar beacon		(MCC)
R/B	Radar beacon (RB is preferred)	RTTV	Real time television
RBA	Recovery beacon antenna (VHF)	RZ	Return to zero
RBE	Radiation biological	R&Z	Range and zero
	effectiveness		
R/C	Radio command	S	ASPO (Apollo Spacecraft
R/C	Radio control		Project Office)
RCC	Range control center	S-	Saturn stage (prefix)
RCC	Recovery control center	SA	RASPO - Atlantic Missile
RCC	Rough combustion cutoff		Range
RCS	Reaction control system	SA	Shaft angle (used by MIT)
RD	Radiation detection	SA	Saturn/Apollo
R&D	Research and development	SACTO	Sacramento test operations
RDMU	Range-drift measuring unit	SAE	Shaft angle encoder
R/E	Re-entry	SAL	San Salvador Island
REG	Regulator		(tracking station)
rem	Roentgen equivalent man	SAL	Supersonic Aerophysics
RES	Restraint system		Laboratory
RF	Radio frequency	SAR	RASPO - Atlantic Missile
RFI	Radio frequency interference		Range (superseded
RG	Rate gyroscope		by SA)
RGA	Rate gyro assembly	SARAH	Search and range homing

SAT	Saturn Systems Office (MSFC)	SCVE	Spacecraft vicinity equipment
SBUE	Switch - Backup entry	SDA	Shaft drive axis
SBX	S-band transponder	SDF	Single degree of freedom
S&C	Stabilization and control	SDG	Strap down gyro
SC	ASPO - CSM (command and service modules)	SDL	Standard distribution list
		SDP	Site data processor
SC	Signal conditioner (SCR is preferred)	SECS	Sequential events control system
S/C	Spacecraft	SED	Space Environment Division (MSC)
SCA	Sequence control area		
SCA	Simulation Control Area (MCC)	SEDD	Systems Evaluation and Development Division
SCAT	Space communication and tracking	SEDR	Service engineering department report
SCATS	Simulation, checkout, and training system (MCC)	SEF	Space Environmental Facility (NASA)
SCC	Simulation control center	SEP	Standard electronic package
SCD	Specification control drawing	SEPS	Service module electrical power system
SCE	Signal conditioning equipment		
SCF	Sequence compatibility firing	SET	Spacecraft elapsed time
SCGSS	Super-critical gas storage system	SF	Static-firing
		SFX	Sound effects
SCIN	Scimitar notch (T/C)	SG	ASPO - G&C (guidance and control)
SCIP	Self-contained instrument package	SGA	RASPO LM - GAEC Bethpage (superseded by SLR)
SCM	ASPO - CSM (superseded by SC)		
SCO	Subcarrier oscillator	SGC	ASPO - G&C (superseded by SG)
S/CO	Spacecraft observer		
SCP	ASPO - CSM administration	SGE	ASPO - G&C engineering
SCPA	SCS control panel	SGP	ASPO - G&C administration
SCR	Subcontractor	SGR	RASPO - G&C MIT, Boston
SCR	Silicon controlled rectifier	SHA	Sidereal hour angle
SCR	Signal conditioner	SHF	Super high frequency
SCR	RASPO CSM - NAA, Downey	SI	ASPO - system integration
SCRA	RASPO CSM - NAA, Downey - administration	SI	Systems integration
		S/I	Systems integration (SI is preferred)
SCRE	RASPO CSM - NAA, Downey - engineering	S-I	Saturn I first stage
SCRR	RASPO CSM - NAA, Downey - reliability	SIA	Systems integration area
		S-IB	Saturn IB first stage
SCS	Stabilization and control system	S-IC	Saturn V first stage
		SID	Space and Information Systems Division (NAA)
SCT	Scanning telescope		
SCT	ASPO - CSM systems test	S&ID	Space and Information Systems Division (NAA)
SCTE	Spacecraft central timing equipment	S-II	Saturn V second stage

SITE	Spacecraft instrumentation test equipment	SPAF	Simulation processor and formatter (MCC)
S-IV	Saturn I second stage	SPDT	Single-pole double-throw
S-IVB	Saturn IB second stage and Saturn V third stage	SPP	ASPO - program plans and control (superseded by SP)
S-IVB G	Launch vehicle guidance system	SPS	Service propulsion system
SL	Star line	SPST	Single-pole single-throw
SL	ASPO - LM (lunar module)	SRD	Spacecraft Research Division (MSC) (superseded by STD)
SLA	Spacecraft LM adapter	SRO	Superintendent of range operations
S/L	Space laboratory	SRS	Simulated remote station (MCC)
SLE	ASPO - LM engineering	SS	ASPO - spacecraft
SLM	ASPO - LM (superseded by SL)	S/S	Samples per second
SLM	Spacecraft laboratory module	S/S	Subsystem
STLOS	Star line-of-sight	SSA	Space suit assembly
SLP	ASPO - LM administration	SSB	Single sideband
SLR	RASPO LM - GAEC Bethpage	SSC	Sensor signal conditioner
SLT	ASPO - LM systems test	SSD	Space Systems Division (USAF)
SLV	Space launch vehicle		
S/M	Service module	SSDF	Space Science Development Facility (NAA)
SMD	System measuring device		
SMJC	Service module jettison controller	SSE	Spacecraft simulation equipment
S/N	Signal-to-noise ratio	SSI	ASPO - systems integration (superseded by SI)
S/N	Serial number		
SNA	RASPO NAA, Downey (superseded by SCR)	SSM	Spacecraft systems monitor
		SSO	Saturn Systems Office (MSFC)
SNAE	RASPO NAA, Downey engineering (superseded by SCRE)	SSR	Support Staff Rooms (NASA)
		SSS	Simulation study series
SNR	Signal-to-noise ratio	SST	Simulated Structural Test (NASA)
S/O	Switchover		
SOC	Simulation operation computer (MCC)	SST	Spacecraft Systems test
		ST	Shock tunnel
SOFAR	Sound fixing and ranging	STC	Spacecraft test conductor
SOM	Suborbital mission	STD	Spacecraft Technology Division (MSC)
SOP	Standard operating procedure		
SP	ASPO - project integration	STMU	Special test and maintenance unit
SP	Static pressure		
SPA	S-band power amplifier	STS	System trouble survey
ACE	Automatic checkout equipment	STU	Static Test Unit (NASA)
		STU	Special test unit

STU	Systems test unit	TRDA	Three-axis rotational control - direct A
SVE	Space Vehicle Electronics (DAC) (superseded by Astrionics Branch)	TRDB	Three-axis rotational control - direct B
SW	Sea water	TRNA	Three-axis rotational control - normal A
SW	RASPO - White Sands Missile Range	TRNB	Three-axis rotational control - normal B
SWT	Supersonic wind tunnel	TTE	Time to event
SXT	Space sextent	TTESP	Test time-event sequencer plan
SYS	System	TTY	Teletype
TACO	Test and checkout station	TV	Television
T/B	Talk back	TVC	Thrust vector control
TBD	To be determined	TVCS	Thrust vector control system
TC	Test conductor	TWT	Transonic wind tunnel
TC	Transfer control	TWX	Teletype wire transmission
TC	Transitional control		
T/C	Telecommunications	UA	Urinalysis
T/C	Thrust chamber	UDL	Up-data link
TCA	Thrust chamber assembly	UDMH	Unsymmetrical dimethyl hydrazine (fuel)
TCA	Transfer control A register	UHF	Ultra high frequency
TCB	Technical Coordination Bulletin (MSFC)	USB	Upper sideband
TCOA	Translational control A	USBE	Unified S-band equipment
TCOB	Translational control B		
TCSC	Trainer control and simulation computer	ΔV	Velocity change (differential)
TD	Technical Directive (MSFC)	VAB	Vehicle assembly building
TDA	Trunnion drive axis	VAC	Volts ac
TDR	Technical data report	ΔVD	Velocity change display
TE	Transearth	VDD	Visual display data
TEC	Transearth coast	VEDS	Vehicle Emergency Detection System (NASA)
TFE	Time from event	VGP	Vehicle ground point
TJM	Tower jettison motor	VHAA	Very high altitude abort
TK HTRS	Tank heaters	VHF	Very high frequency
TLC	Translunar coast	VLF	Vehicle launch facility
TLI	Translunar injection	VLF	Very low frequency
TLS	Telescope	VOX	Voice-operated relay
TM	Telemetry	VRB	VHF recovery beacon
TMG	Thermal micrometeoroid garment	VSC	Vibration safety cutoff
T/M	Telemeter	VTF	Vehicle test facility
TP	Test point	VTS	Vehcile test stand
TPA	Test point ACE		
TPS	Thermal protection system	W/G	Water-glycol
TR	Test request	W-G	Water-glycol
T/R	Transmit/receive		

WGAI	Working Group Agenda Item (MSFC)	WSMR	White Sands Missile Range
WHS	White Sands, New Mexico (remote site)	XCVR	Transceiver
WIS	Wallops Island Station (NASA) (Wallops, Va.)	XDUCER	Transducer
W/M	Words per minute	XEQ	Execute (ACE)
WMS	Waste management system	XMAS	Extended Mission Apollo Simulation (100 days) (NASA)
WODWNY	Western Office, Downey	XMTR	Transmitter
WOM	Woomera, Australia (remote site)	Z	Astronaut Activities Office
WOO	Western Operations Office (NASA)	ZI	Zone of Interior (continental USA)
WPM	Words per minute	ZZB	Zanzibar, Tanganyika (remote site)

SYMBOLS

ΔP	Delta P	ΔV	Delta V

TERMS

ABLATIVE MATERIAL	During entry of spacecraft into the earth's atmosphere at hypersonic speeds, aids in the dissipation of kinetic energy and prevents excessive heating of the main structure.	APOLLO	A term generally used to describe the NASA manned lunar landing program but specifically used to describe the effort devoted to the development test and operation of the space vehicle for long duration, earth orbit, circumlunar, and lunar landing flights.
ABORT	Premature and abrupt termination of a mission because of existing or imminent degradation of mission success probability.	APOLLO SPACE-CRAFT	The vehicle required to perform the Apollo mission after separation of the final launch stage. It consists of the command module (C/M),
APOGEE	A point on the orbit of a body which is at the greatest distance from the center of the earth.		

www.ingramcontent.com/pod-product-compliance
Lightning Source LLC
Chambersburg PA
CBHW050405110426
42812CB00006BA/1804